THE CASE FOR
INTERFERON

THE CASE FOR
INTERFERON

How a 1980s Cancer Drug Might Be the Wonder Therapy for the Twenty-First Century

DR. JOSEPH CUMMINS
and KENT HECKENLIVELY, JD

FOREWORD BY DR. JUDY MIKOVITS

Skyhorse Publishing

Skyhorse Publishing books may be purchased in bulk at special discounts for sales promotion, corporate gifts, fund-raising, or educational purposes. Special editions can also be created to specifications. For details, contact the Special Sales Department, Skyhorse Publishing, 307 West 36th Street, 11th Floor, New York, NY 10018 or info@skyhorsepublishing.com.

Skyhorse® and Skyhorse Publishing® are registered trademarks of Skyhorse Publishing, Inc.®, a Delaware corporation.

Visit our website at www.skyhorsepublishing.com.

10 9 8 7 6 5 4 3 2 1

Library of Congress Cataloging-in-Publication Data is available on file.

Cover design by Kai Texel

Print ISBN: 978-1-5107-6550-4
Ebook ISBN: 978-1-5107-6551-1

Printed in the United States of America

Contents

Foreword by Dr. Judy Mikovits   *vii*

Introduction – The Great Promise of Interferon   *xiii*

Chapter 1 – An Ohio Farm Boy   1

Chapter 2 – The Life of a Vet   11

Chapter 3 – The Russians Are Smart   21

Chapter 4 – Involvement with Interferon   33

Chapter 5 – Interferon Gets Hot   49

Chapter 6 – Interferon in Humans   65

Chapter 7 – A Veterinarian Gets HIV-AIDS   77

Chapter 8 – Interferon Disaster   95

Chapter 9 – The State of Interferon Today   115

Chapter 10 – Interferon and COVID-19?   135

Testimonials   *145*

Endnotes   *149*

Acknowledgments   *159*

Disruptive Ideas Are ALWAYS Fought by the Establishment

By Dr. Judy Mikovits

It's remarkable to realize that interferon marked my first entry into science more than forty years ago, and we're still talking about it. In those decades I've certainly seen a lot, from rampant misogyny to criminal attacks on new and promising therapies, but I think it stems from the same root cause, a fear of change.

In 1980, I was finishing my studies—a major in chemistry and a specialization in biology—at the University of Virginia. This is significant because only forty years ago there was not even a degree program in biochemistry. It was there I encountered my first misogynist dressed in a white lab coat.

An assistant professor voiced his objective to give failing grades to female premed students, as he believed women should not attend medical school. Most of my female classmates simply avoided his classes. But my degree program required three classes taught by this professor. Naive and fueled by my passion for natural products chemistry, I persisted.

My poor grades in two of these classes made it clear by my third year that I would not have the grades to go to medical school. In fact, I struggled to keep the grade point average necessary to keep my scholarships. The last class of my senior year was a biochemistry lab class required for all premed students, of which there were approximately five hundred.

In a lab, I was home. I am a lab rat. For my lab reports, I received the highest grade of all five hundred students. Yet to my horror, the final exam contained not questions about biochemistry labs, but the type of advanced organic chemistry questions that I had failed previously. I received the worst grade on that exam of all five hundred students, and instead of the A+ I had earned, I was given an A- in the class.

I protested to the dean of students, who was sympathetic. But because others had improved their grades granting them entrance into medical school, he could do nothing but reprimand the professor. In the future, he made it a requirement that final exams be given only on the subject matter of the class.

Shouldn't that have been the requirement all along?

I remember the day as if it were yesterday. I almost quit school only weeks before graduation on principle. I thank God for my friend Katie, who talked sense into me as I sat sobbing on the lawn in front of the rotunda wondering what happened to Thomas Jefferson's principles of rational thought and morality.

Apparently, they did not apply to women.

***

On March 31, 1980, I saw the cover of *TIME* magazine. It showed a droplet from a hypodermic needle and the words "INTERFERON – The IF Drug for Cancer."

I knew immediately what I wanted to do.

The National Cancer Institute (NCI) advertised for a protein chemist to purify interferon in the *Washington Post* over Memorial Day weekend only a week after graduation. I applied, and, on June 10, 1980, I started my dream job, purifying interferon at the NCI facility at Fort Detrick in Frederick, Maryland.

Every step of the way I encountered misogyny and corruption. The male chemist I took over for was leaving to attend medical school. After I started, I received a formal letter stating that the technician job would pay only $10,000, rather than the $15,000 given to the previous male chemist. I was told the man had a family and that was why he was paid more for the same job. I vehemently protested, but to no avail.

I kept working anyway because I did not care about money. I was in heaven. I was doing critical scientific work that was helping people. We made the first interferon given to cancer patients, and some were doing very well. We also purified many cancer drugs from plants. I naively thought I'd spend a few years making drugs and maybe then get an opportunity to go to medical school.

***

The events of 1982 would end that dream for me forever as the first human disease-causing retrovirus, HTLV-1, was isolated. My team was assigned to purify very large quantities of the virus, growing the virus in a 200-liter fermenter in the human cell line, HUT-102.

The problem was no one knew anything about the transmission of the virus, and the personnel were not provided any kind of safety precautions or training. Even more alarming to my supervisor and me, several of the female technicians were pregnant. We wrote a formal letter saying we could not complete the job without adequate safety precautions. A few weeks later I received a letter stating that my position had been eliminated. That is, that the NCI no longer needed a protein chemist.

They did not fire me. They simply said my position was part of a force reduction.

***

Devastated but not giving up, I attended a seminar about immune modulator proteins called cytokines. I learned a new program called Biological Response Modifiers was being started to develop cytokines as therapeutics for cancer.

This was the first translational research program in the country, taking what we learned in the lab and quickly transitioning as safely as possible to human trials.

I approached the scientist giving the seminar and asked if the program needed a protein chemist. I naively told him I thought those cytokines might be important and could possibly cure cancer. Fortunately, he did not laugh but invited me to his office to talk. After a wonderful discussion, he told me of an investigator who needed a technician and arranged an interview with Dr. Frank Ruscetti, which would be the start of our longtime collaboration.

***

In that first interview. Frank said, "I see from your résumé that you've purified Interleukin 2 and HTLV-1."

I replied, "Yes."

Frank asked, "So I suppose you read the literature about them?"

"Of course."

"Can you tell me who wrote those papers?"

I shrugged. "Doesn't matter to me who wrote them. It's the science that
matters."

I saw a flicker of anger run across Frank's face. "I did," he said.

"Well that's the end of this job!" I thought, certain I'd blown it.

We proceeded to talk about the Boston Celtics basketball team, as there
was a poster of Bill Russell and Larry Bird on his wall, and I've always been
something of a sports nut.

I left a few minutes later, convinced I'd have to keep looking. However,
fate had different plans for me. Frank must have liked my total commitment
to science because he went to the hiring manager to tell him he wanted me
for the job.

The hiring manger replied that he couldn't hire me because I was a
troublemaker.

He asked why I was a troublemaker.

"She asks too many questions," came the response.

He replied, "She's a scientist. She's supposed to ask questions! You hire
her or I'll have you fired!"

And I was.

That started a decades-long collaboration with Frank, a scientist of total
integrity, an oasis in the middle of a plague of corruption.

***

In 2009, after many long years of building my career, Frank and I published a blockbuster study in the journal *Science*, linking a newly discovered
mouse retrovirus, XMRV (xenotropic murine leukemia virus-related virus),
to chronic fatigue syndrome (also known as ME/CFS, or myalgic enceph-
alomyelitis/chronic fatigue syndrome) and later suggesting it might also be
responsible for many other diseases, such as cancer and autism.

A large body of evidence suggested the zoonosis of these cancer and
neuroimmune disease-causing mouse viruses had occurred via biological
therapies, including vaccines. Especially vaccines against viruses. Scientists
agreed. It is possible that XMRV particles were present in virus stocks

cultured in mice or mouse cells for vaccine production and that the virus was transferred to the human population by vaccination.

While these claims have engendered substantial controversy, the general public cares more about solutions, rather than what may have happened in the past. That's where this fine book by veterinary professor Dr. Joseph Cummins and my frequent coauthor, Kent Heckenlively, can contribute so much to the conversation.

The main argument put forward by Dr. Cummins is remarkably simple. Interferon exists in the body in very tiny amounts during infection from a virus or other pathogen and directs the immune system to successfully repel the intruder. The public health cowboys, who believe that anything that's good in small amounts MUST be better in larger amounts, essentially destroyed the promise of interferon in humans. Interferon is now allowed for widespread use in animals and limited use in humans, but, as Dr. Cummins argues, its full potential has not yet been realized.

It's time to take another look at interferon.

INTRODUCTION

The Great Promise of Interferon

The recent worldwide pandemic of the Wuhan coronavirus (COVID-19) in 2020 and other deficiencies in health care serve as the motivation for me to speak out now, although I have rarely been quiet over the years. I doubt there's a single person in the world who's not been impacted by recent events, from the crashing of financial markets, lockdowns, the quarantining of entire countries, race riots, and the closing of national borders.

These events did not take place by chance. I believe they were the result of foolish, short-sighted decisions, likely by people looking for quick rewards to existing industries, while leaving the future in great peril. I have played a small part in these questions and believe humanity needs to be made aware of research already done on interferon, and how it can be quickly scaled up to deal with some of our most pressing health needs.

The science was so robust for interferon that although it is not used much for humans, it's sometimes used by veterinarians for animals. At the dawn of the 1980s, interferon was looked at as the most promising cancer intervention, landing on the cover of *TIME* magazine on March 31, 1980.

The coronavirus that began in China at the end of 2019 is but a harbinger of greater outbreaks if we do not learn the lessons related to it. And these are not hard lessons to learn because so much of the research has already been done.

You might reasonably ask: What are interferons?

Interferons are small proteins created by all nucleated cells, most likely in response to viruses. Interferons have direct antiviral and anticancer effects and can modulate the activity of the immune system. They are

XIV The Case for Interferon

potent mediators of the immune defense system of people and animals by binding to receptors on the surface of certain cells.

The cover of the March 31, 1980, issue of *TIME* magazine showed an extreme close-up of a syringe with a single reddish-pink drop hanging from the tip and the screaming headline "INTERFERON – The IF Drug for Cancer." The headline was a play on words, as interferon was often shortened in scientific discussions to the letters IF, and scientists were trying to determine "if" interferon was a game-changing treatment for cancer. The article opened with a vivid description of the development of cancer:

> It can start in just one of the body's billions of cells, triggered by a stray bit of radiation, a trace of toxic chemical, perhaps a virus or a random error in the transcription of the cell's genetic message. It can lie dormant for decades before striking, or it can suddenly attack. Once on the move, it divides to form other abnormal cells, outlaws that violated normal genetic restraints. The body's immune system, normally alter to the presence of alien cells, fails to respond properly; its usually formidable defense units refrain from moving in and destroying the intruders.[1]

The *TIME* magazine cover story brought interferon to the attention of the general public, inspiring many researchers, including a young Dr. Judy Mikovits, who was then in her senior year at the University of Virginia and has been kind enough to write the foreword to this book. We know many things can cause cancer, from radiation, to chemicals, to viruses, but we don't really know why.

Aberrant cells are normally destroyed by the body's immune system, but cancer cells somehow inactivate or evade the body's immune response. Some have theorized that cancer cells go "stealth," not alerting the body to their presence, or they broadcast "all clear" signals to the immune system. The question that would rage for more than forty years is whether or not interferon function as a fire alarm or police siren that warns the body that intruders are on the loose.

The article did a good job summarizing how difficult it was to obtain interferon (first identified in 1957), but thanks largely to the dedicated work of a Finnish virologist, Kari Cantrell, scientists were able to start synthesizing larger amounts. The 1980 *TIME* article added:

> Still, researchers now had enough interferon to move studies out of the laboratory and into the clinic. In 1972 virologist Thomas Merigan of Stanford

University, and a group of British researchers began studying IF's effect on the common cold. Soviet doctors were claiming success in warding off respiratory infections with weak sprays of IF made in a Moscow laboratory. Merigan and his colleagues gave 16 volunteers a nasal spray of interferon one day before and three days after they were exposed to common cold viruses. Another 16 volunteers were subjected to the same viruses without any protection. The results seemed miraculous. None of the 16 sprayed subjects developed cold symptoms, but 13 of the unsprayed did. There was one catch: at the IF strengths that Merigan used, each spray cost $700.[2]

There's an old expression about the uneven state of scientific progress: "We can go to the moon, but we can't even cure the common cold."

However, research from the Soviet Union and scientists from Stanford and the United Kingdom were showing a cure for the common cold was tantalizingly within reach. It was just expensive. But if we could get the cost down, who knew what else we might be able to do with it?

Merigan and his Stanford team were ready to move on from the common cold to other conditions, as were several researchers from around the world:

In the years since, Merigan and his Stanford team have successfully used IF to treat shingles and chicken pox in cancer patients. In other studies, IF has prevented the recurrence of cytomegalovirus (CMV), a chronic viral disease that sometimes endangers newborn babies and kidney transplant patients. Israeli doctors have also used IF eyedrops to combat a contagious and incapacitating viral eye infection commonly known as "pink eye." Researchers are now trying a combination of IF and the anti-viral drug ara-A in patients with chronic hepatitis B infections. Interferon investigators have high hopes that the drug will be equally active against other viral diseases.[3]

I hope you're starting to understand why I claim interferon might be a new wonder drug, the way penicillin was a breakthrough therapy for treating bacterial infections. I find myself longing for the way scientists approached their work in the 1970s and 1980s, before the big pharmaceutical companies could exert such a stranglehold on new therapies. Medicine should not be driven solely by the profit motive, but by improvements in human health and longevity. I consider it an abomination that pharmaceutical companies are traded on stock exchanges as commodities, rather than as public utilities for the good of all.

And while today we know some long-term viral infections can lead to cancer, it was less clear in 1981, although it was strongly suspected, as stated in the *TIME* cover story:

> The concept that IF might also be effective against cancer may have occurred spontaneously to several researchers after the work of Isaacs and Lindenmann was confirmed. After all, it had already been shown that some animal cancers were caused by the polyoma virus. Though no human cancer virus has yet been definitely identified, some tumors seem linked to viral infections.[4]

Perhaps the most fanciful part of the article was the first public mention of interferon, which happened in a *Flash Gordon* comic strip from July 12, 1960, written by Dan Barry. In the strip, the crew becomes infected with an extraterrestrial virus when they are in deep space and cannot make it back to Earth in time for treatment.

"This **COULD** be it!" declares the doctor on the spaceship. "**INTERFERON!** It knocked out the virus in the lab animals!"

"**HURRY!**" cries another crew member.

In the next frame, the doctor quickly injects a dose of interferon to the sick patient.

"Well, doctor?" comes the anguished question from a crew member.

"The interferon **WORKS!** The fever is going down!" the doctor proclaims triumphantly.[5]

Please understand, I'm not making any claims that interferon will work against space viruses. But since Elon Musk is getting America back into space and has his sights set on Mars, it might be a good idea for any future astronauts to have a supply of interferon on hand for any Earth viruses that hitch a ride on that rocket.

I think the entire story of interferon, as well as its full potential, has not been fully explored. It is my intention in this book to rekindle an interest in interferon and give it the rigorous testing it did not receive.

***

I like to think of interferons as Mother Nature's own defense system to ensure we stay healthy. One might say I'm interested in working with nature, rather than against it, as many in science have done. Nuclear fission gave us the atomic bomb and radioactive fallout, while nuclear fusion, the same process that powers the Sun, has the potential to give us limitless, clean, inexpensive energy. Pesticides certainly improved food production (at least temporarily) but have caused catastrophic losses in insect populations, as well as disturbing natural food chains. An understanding of organic farming and natural ways to increase food production has the effect of providing us with healthier food, but with far less impact on the natural world. I truly believe in using science for the good of our planet, and it's for this reason I believe it's so important for us to understand how interferons are an untapped but easily available tool to ease both animal and human suffering.

Interferon is found in the body in nanogram (billionths of a gram) quantities, particularly in nasal secretions. Despite this basic biological fact, pharmaceutical companies developed and marketed milligram (thousandths of a gram) doses of interferon for human injection that are at least one thousand to ten thousand times the concentration of naturally occurring interferon. The high doses of injected human interferon, unlike anything found in nature, caused many adverse clinical events.

It is my belief that if we want to achieve good health in humans, we must have humility before nature and seek to duplicate the condition we find in healthy people and animals.

Interferon was not developed for oral administration in human medicine because human interferon given orally could not be detected in the blood of test animals. It was assumed that it had to make it into the blood to be effective. This is false because the kidney quickly and efficiently removes interferon from the blood.

What is the fate of nasal secretions? Most nasal secretions trickle down the throat, past the tonsils and other lymphoid cells in the pharynx. Any interferon in the nasal secretions has the opportunity to contact these lymphoid cells described as Waldeyer's ring.

The expression of thousands of genes was modified in cattle by introducing low-dose oral human interferon alpha into their systems. For the purposes of this book, I will always refer to human interferon when I mean human interferon alpha. There are other types of interferon, but unless I specifically refer to them by a different name, I am discussing human interferon alpha, and I am usually referring to low-dose human interferon alpha.

In April 1981, the *New York Times* magazine did a large article on interferon that laid out the differing sides of what had become a raging controversy. It's remarkable to read this article nearly four decades later and find that the two sides of this argument remain dug in at approximately the same place today as they were at that time. If anything, I believe my side has generally lost the argument in the scientific arena as far as using interferon for human health, although the war still rages. As the *New York Times* magazine informed its readers:

> Interferon is a sinuous, sticky protein. Virtually all human and animal cells are capable of manufacturing it, yet they do so only rarely and then in such minute quantities that detection takes enormous laboratory effort. It is so rare that a pound of purified interferon was once estimated to be worth $20 billion. Blood from 270 donors is needed to produce enough interferon to treat one person for a few weeks. The protein's existence has been known for 24 years, but it has been so hard to come by during that time that research has been difficult, if not impossible.[6]

Yes, you might say in the 1980s there was something of a "gold rush" for interferon, and yet like all rushes there was also likely to be a crash, since technology would quickly be able to lower the price of the product to a level below that of most pharmaceutical drugs. One of the most frustrating problems I've encountered in medicine is that the system is not set up to reward innovation. Instead, the rewards flow to those products that create a continuing cash flow for the pharmaceutical companies.

I've heard it claimed that there's no money for pharmaceutical companies in dead people or healthy people. The real money is in the sick but functioning people, who need a daily dose of some medication. Bad systems produce bad outcomes. This system needs to change. Should there be some sort of government incentive program for the company that creates a medication that can, say, cure cancer, diabetes, or Alzheimer's disease?

Maybe. They've worked in practically every other situation involving human nature.

I have no interest in bankrupting the pharmaceutical companies. But they need to deliver health, not a lingering dependence on a pharmaceutical drug. The *New York Times* magazine article continued with its summary of the state of interferon in 1981:

> But now, although interferon in recent months has swung abruptly back and forth on the pendulum of scientific promise and pessimism, many important medical researchers and physicians regard it as having greater potential as a cancer treatment than anything else to come on the scene in the last 10 years.

These hopes are counter-balanced by interferon's scarcity. But to the extent that is has been available for use as part of research and testing, it has been keeping people alive.

Preliminary and necessarily incomplete experiments show, in the opinion of many reputable scientists, that interferon is effective in shrinking some types of tumors and in stopping their spread. Moreover, it has done so without the terrible side effects that cause many people to dread chemotherapy and radiation. Beyond this, scientists believe, interferon has startling capabilities for fighting viruses, including what may prove to be, under certain controlled conditions, the first real cure for the common cold.[7]

In this book you will learn about the back-and-forth of the scientific establishment on interferon. Let me be clear on what I believe.

Low-dose oral interferon most closely mimics how a healthy body responds to a virus or pathogen, or some other conditions in the body.

But most of the scientific establishment, utilizing some wobbly form of a cowboy slash-and-burn strategy, where if a little is good, more must be better, ramp up the dosage of interferon to a level nature never intended.

Our bodies are designed by nature to be healthy. The job of a scientist, therefore, is to discover what is going right in the body of a healthy person, and how that differs from what is going wrong in the body of a sick person. Our wonderful technological tools give us the opportunity to find nature's hidden path to health. But if we do not read the clues correctly, we will go down the wrong road. What is the origin of cancer in our bodies? I think the *New York Times* magazine article had it right in 1981:

Our own bodies, not the direct intervention of foreign agents, create cancers, and the hope is that interferon, a natural product, can be relied upon to get the body to stop destroying itself. It is believed that interferon somehow recharges the body's own immunology system, attracting the body's natural "killer" cells to the cancerous territory, where they conquer the malignant cells. Cancer seems to overpower these killer cells, but interferon returns their vitality and effectiveness.[8]

While the notion of our own bodies causing cancer may seem hard to understand, it might also bring up common questions. Doesn't smoking cause cancer? That's partly true. Smoking basically destroys part of your immune system's ability to respond, leading to the development of cancer.

When one considers the number of chemicals in our environment and then our lack of understanding of all their possible interactions on our health, one begins to understand how blind we are about cancer development. We lack critical information to make good decisions. But if we have an answer which could counteract this damage, like interferon, that would be a remarkable development.

However, there can be some side effects of introducing artificial interferon into a person's system:

> The effect felt most clearly by some interferon recipients is a lethargic condition that has been dubbed "interferon malaise." After one week of interferon treatments, many patients lose energy: They don't want to eat; they don't want to do anything. They flop down for long, numbing naps. They get depressed easily.
>
> "I was very tired," says Edward Tyler, "just dog-tired all the time. After I took my shot in the morning, I just didn't have the energy to do anything. I couldn't walk with a brisk pace like I used to. So there's no question that interferon has side effects for me. But in comparison to some of the other stuff I've been through, it was nothing."[9]

What I've found in my veterinary practice (and when I've collaborated with medical doctors) is that it's usually best for individuals to take interferon for five days, usually Monday through Friday, and then take an interferon "holiday" on Saturday and Sunday. That will normally handle any potential problem with any malaise.

Like any medication, for animals or humans, dosage is key.

At one dosage, a particular substance can be a cure, but at another level, a poison. I also believe there are some unique characteristics to interferon. Interferon does NOT exist at high dosages in healthy biological organisms. So why would we think that high dosages will lead to good results? Even in 1981, this pattern was becoming clear, according to the *New York Times* magazine article:

> The key to this situation seems to be dosage. A recent study subjected mice to a high dosage of interferon from their day of birth onward. The creatures died on their eighth day. A control group given 10 times less interferon suffered no such fate. The reason high doses of interferon may have destructive effects remains unclear, but researchers find the problem worrisome.[10]

"Worrisome?"

Did I really read a sentence in which researchers said they found death to be a worrisome problem?

I was trained to believe the death of a subject or patient represented a complete failure.

The article ended with what I think may be some of the most promising aspects of interferon and how it can be a useful tool in our arsenal against disease:

> And while cancer fighting has been grabbing all the headlines, interferon's potential as a protection against viruses remains powerful. The protein blocks the action of every virus it has been tested against in the lab. It seems to be able to stop some common cold viruses dead in their tracks, and to offer hope for eliminating the dangerous carrier of chronic viral hepatitis. However, more testing must be done in these areas, too, before interferon's antivirus potency can be fully judged."

One can easily understand from the *New York Times* magazine article why so many researchers like me were excited by the possibility of interferon.

But before I tell you more about interferon, I'd like to first describe how I became one of the world's foremost advocates for its widespread use.

CHAPTER 1

An Ohio Farm Boy

My father joined the US Navy in 1935 when he and some friends drove to Cincinnati to visit a Navy recruiter. It was the depths of the Great Depression, and he considered himself lucky to be able to enlist and have a job.

He didn't consider himself lucky several years later when he and my mother were at Pearl Harbor, Hawaii, when the Japanese attacked on December 7, 1941. Nonessential personnel were evacuated from Hawaii, and I came into this world a few months later in the sunny climate of Southern California in a nation at war.

My father remained at Pearl Harbor to help clean up the devastation and get the country ready for a long global fight. After the attack on Pearl Harbor and the entry of the United States into WWII, Dad rose in rank quickly and was soon a chief petty officer, boiler tender.

In the war, two ships upon which he served were torpedoed by the Japanese and sunk. After the sinking of one of those ships, Dad spent what he euphemistically called "a night in the water," before being rescued. What worried my father the most? Drowning? Sharks? Maybe being strafed by Japanese planes? He never said specifically, but he told me that whenever he went to sleep, he dreamed he was "back in the water." He felt lucky to have survived the war, but there were invisible scars he carried for the rest of his life.

Dad was determined to raise his sons on a dairy farm in Ohio, far from the ocean. When my father retired from the US Navy in 1955, the family moved to Logan County, Ohio. There are few jobs as demanding as being a dairy farmer. Cows need attention every day, all year long. The milking had to occur every twelve hours. There were no holidays. And I think it was due to the focus on the livestock on that farm that I first developed my deep affection for animals.

Dad was determined that I learn to work hard, and there was no better place to instill a work ethic than the dairy farm. I never overcame the feeling that, at day's end, there was still unfinished work to do. Just like other farm boys, I could put a hundred bales of hay or straw on a wagon so none would fall off. If necessary, I could back two full wagons, hooked together, into the barn. I was a good, hard-working, tough farmhand. The backbreaking work of farm life made everything else pale in comparison. As far as I was concerned, anything other than working on a farm was a vacation, sunshine and lollipops all day.

I was trained to work hard and put in long hours. As a veterinarian, I was busy all day and at night, if there was an emergency. I was ready to dash from my house to the office at a moment's notice to lend assistance. Later, when I became a professor of veterinary medicine, I learned there is no end to the hours that can be spent trying to keep abreast of new scientific publications and developments.

I put a terrible burden on my home life by working endlessly. It is one of my great personal failings that I never learned to relax and enjoy anything except work. Even though I have various academic distinctions and publications in the top journals of my profession, I still consider myself a simple Ohio farm boy at heart.

***

I graduated from Washington Local High School in Lewistown, Ohio, best known as the place where Walter Alston started his legendary baseball coaching and managing career. His example of quiet, dogged persistence stood as an example to all of us.

Alston was nicknamed "Smokey," playing nineteen years in the minor leagues and only appearing in a single major league baseball game for the St. Louis Cardinals in 1936. In 1954, he was captain of the Nashua Dodgers (the farm team for the Brooklyn Dodgers, and the first integrated baseball team in the United States).

But it was as a major league manager that Alston really made his mark. He was often referred to as "the Quiet Man," for his calm demeanor, managing the Brooklyn, then Los Angeles, Dodgers for twenty-three years, from

1954 to 1976. In those years, his Brooklyn team won the World Series in 1955, and the Dodgers had more than two thousand wins during his tenure as manager. In Los Angeles, his team won three World Series, in 1959, 1963, and 1965; he managed seven National League All-Star teams to victory; and he was selected Manager of the Year six times. Alston's dogged determination, "never-say-die" attitude, and calm professionalism loomed large as an example for all of us at his former high school.

Alston was well into his legendary managing career with the Dodgers by the time I graduated from high school in 1960 and enrolled in fall classes at The Ohio State University. I was accepted into the College of Veterinary Medicine after two years of undergraduate work in the College of Agriculture. Having been raised on a dairy farm, it was my intention to pursue a career as a practicing veterinarian.

Little did I realize my intentions were far removed from what eventually transpired. I had set my sights too low for the future that awaited me, when my discoveries would cause heated controversies in countries across the globe like Kenya during the middle of the HIV-AIDS epidemic.

I attended the College of Veterinary Medicine from 1962 to 1966. The competition in the professional curriculum was more intense than anything I'd previously experienced. It wasn't much fun. As one of the youngest members of the class, I had trouble adjusting to the rigors of the schedule and subject matter. I was particularly terrified by embryology, the study of how a fetus develops to maturity prior to birth. Embryology seemed like a foreign language to me, as every descriptive term was new, and the learning pace was fast. I agonized over identifying the embryonic organs and tissues on our class slides, and it took the whole first quarter before I was the least bit comfortable with the subject matter.

Anatomy was the most boring class I'd ever experienced. Our professor, a wonderful man but a terrible speaker, started his anatomy lectures each morning at 8 a.m. in a quiet, gentle monotone.

Unfortunately, I was not a coffee drinker at the time.

Coffee probably would've helped.

The class was a struggle for everyone to keep awake, as each artery, nerve, bone, and muscle was discussed and described ad nauseam. As if all of that weren't enough, late in the first quarter of my veterinary education, I developed acute appendicitis.

On a Friday, I left a bacteriology lecture and checked into the student clinic. A quick examination and blood test confirmed a diagnosis of appendicitis, and I had surgery that Friday night. By Tuesday, I was back in class, but I'd missed an important anatomy test. I was called into the professor's office for a verbal makeup test. I did not expect to do well.

However, before the exam, he asked me where I was from. I told him I was from Bellefontaine, Ohio, in Logan County. He checked my name and my father's name and concluded he'd known my great-grandfather. He told me that, as a boy, he'd often ridden in a horse and buggy with my great-grandfather, and my great-grandfather had once taken him to the local county fair. He remembered my great-grandfather with much affection and admiration. In looking back, that probably saved my veterinary career.

After briefly reminiscing, he asked a few simple anatomy questions and dismissed me.

I got a "B" for the course but probably deserved to fail.

My great-grandfather, through his kindness to a boy forty years earlier, had unknowingly helped me survive my first quarter in veterinary college. What are the chances of such luck? I felt like some sort of divine providence was guiding my life and didn't want to waste the opportunity I'd been given.

After a shaky first quarter, I applied myself even harder in the second quarter and started to see results, suggesting I might even have a talent for this kind of study. I set a record-high test score on a practical histology test in which we moved from microscope to microscope identifying tissues.

In early 1963, we studied large and small lymphocytes, macrophages, and plasma cells and heard speculation about their role in the overall immune response. I was fascinated by it all. The professors felt science was at a turning point in understanding both animal and human health, and many of us shared that enthusiasm.

I did not know at the time how important these cells would become in my future as a research scientist studying immunology, virology, and interferon.

***

I was in my junior year of the College of Veterinary Medicine in 1965 when the military came calling. Because of the war in Vietnam, practicing veterinarians were being drafted out of practice to serve in the US Army.

My father had spent twenty years in the US Navy before retiring to an Ohio farm, so I decided it was possible that I might enjoy a career in the military.

Dr. Scothorn was teaching veterinary parasitology at Ohio State after a career in which he'd attained the rank of colonel in General Douglas MacArthur's army. After arriving at Ohio State, Scothorn actively recruited veterinary students for the US Army, and I became one of them. He convinced me that if I signed up for three years (instead of the regular two years), he could obtain a juicy assignment for me in laboratory animal medicine at the Presidio in San Francisco.

I volunteered for three years. It was an understatement to say I was surprised when I subsequently received orders to report to Bayonne, New Jersey. My assignment would be to supervise inspections of fruits and vegetables shipped to our troops in Europe.

I took my new orders to Dr. Scothorn and asked how it was possible for him to promise me a laboratory animal medicine assignment in San Francisco and for me to receive, instead, an assignment to inspect fruits and vegetables in New Jersey.

He looked at my assignment for a long time and then said, "Well, I promised you a coast, and I got you a coast." I think this was my first significant experience with the promises of people in power, only to be disappointed in what they actually delivered.

Unfortunately, it was not to be my last.

Later I heard of a case of an US Army physician who sued the US Army because he'd been promised a specific special assignment but had received an undesirable duty station instead. The court was unsympathetic and ruled in the US Army's favor. I remember the judge said that the physician was naive.

"Everyone knows," the judge said, "that US Army recruiters lie."

Wow.

What a thing to tell young people who want to serve their country. I wonder if a soldier could do the same in reverse. *I know I said I promised to charge the enemy, but I changed my mind and I think I'll head to the canteen instead because I'm hungry.*

I think one would get court-martialed.

I spent thirty-nine months (thirty-nine months, six days, four hours, and twenty minutes—but who's counting when you're having fun?) in the US Army. In retrospect, I can categorically state I never did anything of the slightest importance toward the war effort. Later I learned that only 1 percent of veterinarians reenlisted in the US Army Veterinary Corp. The 1 percent reenlistment rate was the lowest of any division in the US military.

I took basic training in San Antonio, Texas, at Ft. Sam Houston and then completed US Army Veterinary Corp training in Chicago at Ft. Sheridan, before I finally arrived in New York for my duty. I lived on Staten Island (Ft. Wadsworth) and worked in Bayonne, New Jersey, for about a year supervising inspection of fruits and vegetables and inspecting food establishments in New Jersey, which provided food to the military. I inspected manufacturing facilities that produced mayonnaise, pickles, bread, cakes, olives, maraschino cherries, ice cream, milk, and other dairy products, as well as pizza, to name a few. I made sure there were doors on restrooms, screens on windows, and some effort at cleanliness.

Overall, I inspected food establishments as best I could, considering I had no confidence in what I was doing and no conviction my inspections were worthwhile. Maybe somewhere in the military there's somebody who can explain to me why the skills of a veterinarian are best suited for work as a food inspector. But I never heard the explanation.

Fortunately, after a year, the post veterinarian left Ft. Wadsworth, and I replaced him. I was so relieved to leave behind the job of food inspector and return to the care of animals, which was my passion. For the two years remaining in my tour of duty, I was the post veterinarian for Ft. Wadsworth and Ft. Hamilton (Brooklyn). These military forts are at each end of the Verrazano-Narrows Bridge.

As a post veterinarian, I held clinics for pets of military dependents on Governor's Island, Floyd Bennett Field (Naval Air Station), and Ft. Wadsworth/Hamilton. In addition, I offered veterinary care to guard dogs at the various missile sites around New York City including Ft. Totten, Ft. Tilden, Highland Park (NJ), Franklin Lakes (NJ), and Suffern (NY). It was a pleasure to visit these missile sites and to care for these magnificent German Shepherd guard dogs.

The most vivid memory of my days working on guard dogs involved a powerful male German Shepherd named "Rocky" who'd developed a habit of licking his left front paw. The habit became destructive, as he would lick his paw until it was bleeding and raw. I would bandage the leg and it would heal, but Rocky would start licking it again once the bandage was removed, or after he chewed off the bandage.

I finally removed him from the missile site and hospitalized Rocky to give him constant attention. I eventually put a cast on his leg to keep him from getting at his paw. He was a very friendly, lovable dog and all military personnel lavished attention and affection on him.

One morning we came to work to find that Rocky had chewed off the cast and *eaten* his paw. His paw was gone. We all had become attached to Rocky and were sickened by the sight of his self-mutilation. I euthanized Rocky and reported his loss to the US Army. Later I talked to other veterinarians who had seen such self-mutilation in dogs in their practices.

Years later, I put human interferon in the drinking water of parrots, cockatoos, and other exotic birds living in a large outdoor enclosure. The purpose of the added human interferon was to offer some protection against bird flu.

Much to our surprise, the birds stopped feather picking, a classic example of obsessive-compulsive disorder. A colleague, John Chamberlain of *Cocky Smart* (yes, that's the real name) in Perth, Australia, also reported cessation of feather picking in his cockatoos given low-dose oral bovine interferon.

This book contains many examples of oral interferon being safe and efficacious in helping manage many diseases in mammals. Our example in birds is not unique in that publications from China, Pakistan, and the United States report that low-dose oral chicken interferon is safe and efficacious in helping manage viral diseases in poultry.

Looking back, I wonder if low-dose human interferon might have saved Rocky's paw.

Feather picking is an animal model for a human disease called trichotillomania, characterized in its worst form by the removal of all body hair including eyebrows and eyelashes. I have not yet had a chance to offer low-dose human or bovine interferon to anyone with trichotillomania, but I believe it would be an interesting and safe experiment to try.

At that time, most of the German Shepherds in the US Army were adopted. I examined prospective dogs and radiographed their hips to rule out hip dysplasia in dogs from the metropolitan New Jersey-New York area. People often bought a cute little puppy but lost interest and gained fear when the puppy grew to eighty or ninety pounds and demonstrated normal canine aggression. If the dogs were selected for the US Army's sentry or guard dog programs, they were sent to Lackland AFB in San Antonio, Texas, for extensive training. It was estimated that, in the late 1960s, even though the dogs were donated, the training and care of the dogs resulted in a cost to the military of five to ten thousand dollars per dog.

One day, a powerful young German Shepherd male was presented to me for examination. The dog was normal in every respect except when I opened the dog's mouth. Dangling from a thin thread of tissue was the dog's right tonsil. Someone apparently had jammed a stick or some sharp object down the dog's throat. I felt surprise and anger that anyone would mistreat a dog. It was the first time I had seen evidence of such horrific abuse. I wished I could jam a stick down the owner's throat. (Sorry, I tend to get very passionate about the mistreatment of animals.)

Early in the Vietnam War, we were told that two thousand German Shepherd guard and sentry dogs were trained, shipped to Vietnam, and placed with the South Vietnamese army. These dogs were promptly marched into the jungle and eaten by the soldiers!

"Military intelligence" had triumphed once again!

***

While in the US Army, I was one of many veterinarians who took on the task of inspecting the nation's fallout shelters. In every major city in those days, fallout shelter supplies were stockpiled in case of nuclear attack. Food, water, radiation detection equipment, and medical and sanitary supplies were standard items in the fallout shelters. The US Army Veterinary Corp was chosen to inspect and report on the status of these supplies after many years of storage.

The "food" consisted of a specially designed nutritious cracker. Research had determined that Americans would not eat a cracker that, during storage, deteriorated into more than three pieces. Therefore, an especially tough, indestructible cracker was manufactured for the nation's fallout shelters. The drinking water in the fallout shelters was in plastic bags inside metal drums holding fifty gallons each. The radiation detection equipment was battery-operated, and most of the batteries were past expiration and useless. The medical supplies were outdated.

The sanitation supplies (toilet paper, sanitary napkins, and toilet seats to put onto empty water barrels) were okay unless the fallout shelter was located at a school, in which case, the sanitary napkins had usually been removed, presumably by teenage girls. I often wondered if the true purpose of these fallout shelters was to provide some measure of psychological comfort, rather than actual protection in case of a nuclear attack.

When I went to the Brooklyn Navy Yard to inspect the fallout shelter there, I was greeted by a US Navy Lieutenant who volunteered ten Navy prisoners to help count the many water drums and inventory the medical supplies, etc. I accepted their help and started the inspection. Unfortunately, two naval prisoners took a bottle of phenobarbital when my sergeant's attention was diverted. Later that evening, the sailors swallowed potentially lethal dosages of the barbiturate. They were spotted weaving and staggering and were hospitalized.

After their stomachs were pumped, the two men survived.

However, my reputation did not.

Army command decided it was *my fault* these men nearly died.

Consequently, I was not allowed to inspect any more fallout shelters.

After Congress was apprised of the condition of the supplies in the fallout shelters, a decision was made to dispose of them. The crackers were still judged to be nutritious, but not especially appealing. I heard the notion of donating these crackers to the inhabitants of some distant starving country was discussed but discarded when it was decided that it would be an insult, since the crackers were considered unsuitable for Americans to eat.

The crackers were then dumped at sea to become fish food.

My army career provided me many opportunities to observe foolish government decisions up close. As I entered my private veterinary practice, I would observe many further misguided government policies, which affect not only the health of animals, but have also effectively denied our human population a very important medical therapy.

CHAPTER 2

The Life of a Vet

While stationed at Ft. Wadsworth on Staten Island, New York, from 1968 to 1969, and after obtaining a veterinary medicine practice license in New Jersey, I began to "moonlight" for a New Jersey veterinarian with a busy practice. Dr. Jones (a pseudonym) was uncomfortable operating and would save some of his surgeries, such as spays and castrations, for me to perform on Saturdays. I'd go into his busy animal hospital each Saturday and operate while he saw clients for their pets' medical care.

One day a spay procedure turned into more than a routine operation when I discovered an ovarian tumor the size of a softball instead of the normal "thumb"-size ovary. Dr. Jones came into surgery and watched me struggle to successfully remove the tumor. He seemed quite impressed with my surgery skills (which, in truth, were quite limited at the time) and accepted me into his practice wholeheartedly from that point on.

It's a good thing the ovarian tumor incident occurred because I possessed no skill or experience in handling clients. He liked to play the role of my mentor and enjoyed showing me how he manipulated and satisfied his clients. Several examples stand out in my memory.

A matronly woman and her twenty-something daughter brought in an eight-year-old Siamese cat to the clinic. I'd just finished a surgery and was running one of the exam rooms to which the cat and owners were admitted.

"I've brought kitty in for a checkup," the older woman said. "Last week Dr. Jones treated kitty for a virus infection."

I looked at the clinic pet record card and could not find evidence of her visit the previous week. "Well, let me start by taking kitty's temperature" was my reply.

As I tried to insert the thermometer into the cat's rectum, the Siamese growled and clawed its way onto the old woman's shoulder.

I stepped back, somewhat surprised, and asked, "Wow, how long has the cat been acting like that?"

The older lady smiled and said, "Eight years."

I then made a classic blunder.

One should *never* criticize a client's pet. But I proceeded to do so by saying, "Too bad you didn't discipline that cat when it was younger."

As soon as those stupid words were out of my mouth, the younger woman said, "See mom, I told you."

By now the older woman knew I was an idiot and said haughtily, "Dr. Jones *never* has trouble with my kitty."

There was no way I could recover from my blunder, so I told her I'd get Dr. Jones from the other exam room. I went to the other examination room and sheepishly told Dr. Jones of my mistake. "Watch how I handle this, Joe," Dr. Jones said.

He entered my exam room and spoke a friendly greeting to the mother and daughter. Then to my surprise, he asked them both to leave the exam room while he gave their cat a shot.

As soon as they departed and he closed the exam room door, he grabbed the cat by the fur on the neck and on the back and slammed the cat down on the exam table.

Bang! Bang! Bang!

The cat hit the table hard, three times.

I think the cat was as surprised as I was.

In fact, I think that was the first time I ever saw a cat act surprised. The cat's head moved from side to side as if to ask, "What's going on?" While the cat was trying to figure out what was happening, Dr. Jones gave it a quick injection of something he grabbed from the table.

He then opened the door, and both women reentered the exam room. The older woman looked right at me and said, "Thank you Dr. Jones, I know my kitty loves you."

***

On another occasion, I was running an exam room when a man phoned and told me his dog was exhibiting severe bloody diarrhea. I checked the clinic pet record card and saw that the dog had been previously treated for hookworms. In those days, a drug (DNP) was injected into dogs to treat hookworms, at a cost to the vet of about $1.50.

The owner told me over the phone that he just did not have the money to undergo another expensive ($14.00) treatment for hookworms. He wanted to know what he could do at home. I put the man on hold and related my conversation to Dr. Jones.

I asked Dr. Jones to take the call, and he wasted no time taking the phone from me and yelling into it, "Bring that dog in here right now!"

Then he hung up.

To my surprise, the man arrived at the clinic in person with the dog and told me the story over again. I took the dog's temperature and smeared some bloody feces on a slide and identified hookworm eggs on the microscope slide. I reported this information to Dr. Jones, who then said, "Watch how I handle this, Joe."

He went into the exam room, looked the owner in the eye, and said, "I hate hookworms. They are a terrible pest. I hate them so much that I have been treating your dog at *my* cost. Now you are going to pay for another treatment because your dog needs it. Is that understood?"

The owner's defiance melted before my eyes. "Oh, Dr. Jones, I had no idea. Thank you, Dr. Jones. Thank you. Oh yes, by all means, treat the dog!" He reached into his wallet and paid $14.00 immediately.

***

Dr. Jones was sneaky at times, and he was not against an occasional placebo treatment. One day an Afghan hound was presented to me, and the owner said, "I'm here for another shot to make my dog's hair grow better." I checked the client record card and could not find any record of shots for the dog.

I went to Dr. Jones and asked him for an explanation. "Oh," he said, "I was giving that dog testosterone shots but was afraid of giving him too much, so I just give him shots of sterile water."

I went back to the exam room and asked the owner how well the shots were working. "They're great," he said. "The dog has never had a better coat."

***

Coincidentally, I left the US Army soon after the 1969 Mets won the pennant. I was in Shea Stadium the night the Mets beat the St. Louis Cardinals and clinched the pennant. The fans were quite happy and excited, and being there is one of the more pleasant memories of my time in New York City. For years, I carried a key chain souvenir given out that night.

I gladly left the US Army when my term ended and headed to Cincinnati, where I worked for a veterinarian who owned two veterinary clinics. I ran one clinic while the veterinarian owner ran the other. I remember my first emergency call in that practice.

A woman called me at eleven at night to complain that her dog was "burning up" with fever. I asked her what the dog's temperature was and she said she didn't know, but the dog "felt hot."

I said, "I don't want to charge you an emergency fee if it's not absolutely necessary. So, please take the dog's temperature and call me back in five minutes."

She called back and said, "I can't take this dog's temperature because the dog keeps spitting out the thermometer." This client obviously didn't know much about her dog, and I had failed to appreciate that. There were no further calls, so I hoped that the dog was fine.

Aroused from a deep sleep one morning at 4:00 a.m., I fumbled for the ringing phone. As the on-call veterinarian, I answered, "Hello, I'm Dr. Cummins" with as much authority as I could muster at that early hour.

A woman on the line said excitedly, "Oh, Doctor, I've got an emergency. My dog won't stop coughing!"

This seemed like a strange emergency, so I asked, "How long has the dog been coughing?"

"Three weeks," was the answer.

Puzzled, I naturally asked, "If your dog has been coughing for three weeks, why are you calling now at 4:00 a.m.?"

"Well," she said, "I'm awake with my sick daughter, and I didn't want to call my pediatrician at this hour, so, I thought I'd call you."

I then discussed the coughing dog and what the owner could do about it.

I should have set my alarm clock for the next day at 4:00 a.m., then called the owner and asked, "How's your dog doing?"

***

One of my most unpleasant memories of practice in Cincinnati occurred with a Dalmatian dog named Dylan. "Oh," I said, "is the dog named for Bob Dylan?"

"No," said the disgusted owner, "the dog is named after Dylan Thomas, the poet."

I should have known I was in trouble.

Dylan was a large Dalmatian with a long tail. The owner had shut a door on Dylan's tail and the dog had started to frequently lick the injured area of the tail, a behavior that irritated the owner. I aggressively treated Dylan to try and break him of his tail-licking habit.

The owner decided to leave town one weekend, and I boarded Dylan at my clinic. Because Dylan was so large, instead of a cage, I decided to give him the run of my office and tranquilized him before I left for the evening.

I returned early the next morning to find blood on the exam room walls and on Dylan's face. My day was ruined when I realized that Dylan's tail was gone—he had eaten his long tail down to a short stub of about four inches. More than twelve inches of his tail was gone!

I anesthetized Dylan and trimmed his tail into a neat "bob" and stitched the end. When the owner returned, she walked into the office with a nice smile and said, "How's Dylan?"

I replied, "You are not going to believe this, but your dog ate his tail." She did not believe me and sought the services of an attorney to see if she could sue me. Maybe I should have come up with some creative lie in the way Dr. Jones had instructed me. But I'm just not built that way. I'm not saying I'm always right, but the artifice of deception does not come easily to me.

Eventually, she forgave me without a lawsuit.

***

One spring day, a man brought his dog (a pug mixed breed) to me with fishing line wrapped around the dog's neck. The business end of the fishing line was anchored in the dog's esophagus, as a radiograph clearly showed. It seems the man was fishing on the Ohio River and pulled his line out of the water. The dog was sitting with his master as the hook and bait landed in the boat near him, and the pug immediately swallowed it whole in one gulp.

The owner was quick-thinking and had grabbed the line and wrapped it around the dog's neck to keep the hook from going farther and entering the stomach. I'd never seen such a case before and did not know what to do. I discussed ripping it out of the esophagus (a dumb idea) or surgically removing the hook.

The owner, far smarter than I, proposed an alternative solution.

"Why not," he suggested, "run the fishing line into a garden hose, then push the garden hose carefully down the dog's throat until the end of the hose reaches the hook. The hose might gently push the hook out of the esophagus and allow extraction of the fishing hook without cutting into the esophagus."

After light anesthesia, I did exactly as the owner suggested, and the hook was removed from the esophagus without incident. I forgot to ask the owner if such an event had occurred before, or if his practical solution was an original thought on his part. Over the years, I've never had another chance to try this procedure.

I always tell that story to drive home the point that many people have good ideas, not just the experts.

***

When I was in practice in Dayton, Ohio, a woman about forty-five years old brought in a male poodle and nervously confided that she suspected her daughter-in-law of sexually abusing the dog. She wanted me to confirm that the dog was sexually abused!

Wow, I could only imagine the tension in that household!

It seems the son was in the service in Vietnam, and the unwelcome daughter-in-law was living with her mother-in-law.

"How was the dog abused?" I asked.

The mother-in-law explained, "She takes the dog upstairs and it seems normal, but the dog comes downstairs nervous and excited."

I wasn't about to get involved in the middle of such a controversy. After examination, I assured the lady that the dog looked normal in all respects and I doubted that sexual abuse was occurring. I didn't want to get sued, so I took myself out of the controversy.

I've never had much use for lawyers, and later events in my life would only sharpen this dislike.

***

In another dog case, I did get involved in a lawsuit and unfortunately testified on the witness stand. A client and friend of mine had a female German

Shorthaired Pointer, which had been so successful in hunting field trials that the dog was a Grand Champion.

When the Pointer became pregnant, the owner accepted the offer of a physician at the Wright Patterson Air Force Base. The doctor offered to care for the dog in his new kennel and to whelp the puppies, in exchange for one of the pups. My client, who lived on a farm north of New Carlisle, Ohio, accepted the offer because the new kennel facilities were far superior to his own kennel.

A week before the puppies were due, the owner called to tell me that the dog was dead. "Would you perform an autopsy?" he asked. I examined the dead dog and my first question was "Do you intend to sue the physician over the death of your dog?"

"No," he said, "I just want to know what killed the dog."

I opened the dog's abdomen and noted a large uterus in which I could palpate ten dead puppies. I then opened the dog's chest and looked at the heart and lungs. Everything looked normal until I looked at the dog's mouth. There, at the commissures of the lips, was a lesion that I guessed was an electrical burn.

When I pointed to the lesion, the owner told me that when he first saw the doctor's kennel, an electrical cord was on the ground running to a radio. It was apparently being listened to by a teenage son who was cleaning the kennel. At that time, my client pointed to the electrical cord and told the doctor that it would have to be removed. The radio and electrical cord were then immediately taken from the kennel. Now, with a dead dog and an apparent electrical burn, I asked the owner again, "Are you going to sue?"

He answered again, "No, I just want to know what killed the dog."

I cut a sample of the burn from the dog's mouth, removed a piece of lung and liver, and sent the tissues to a laboratory in Dayton, Ohio, operated by the Montgomery County coroner. He sent back a report stating that the dog had died of "electrocution."

Now my client wanted to sue.

I was mortified because I had not done a complete necropsy. I had not examined the brain and every organ. I had taken my client's word that he did not want to sue.

As the case wound its way to trial, I was eventually called to testify.

I was afraid the defense attorney would ask about the dog's brain. Could the dog have had a brain tumor? I did not know. I didn't look.

There were so many questions that could have been asked because dogs can die for a myriad of reasons. I'd looked for the most likely cause of death based on my initial observations and conversation with the owner. However, I was certain a good defense attorney could pick apart my relatively cursory work. Isn't that what defense attorneys do in all those TV crime shows? Make the witnesses look like fools?

Had I known a lawsuit was going to stem from this case, I would have advised the owner to take the dead dog for a complete and thorough autopsy at The Ohio State University, College of Veterinary Medicine, in Columbus.

But my fears were unfounded.

The defense attorney didn't know which questions to ask. He could have embarrassed me and discredited my testimony. But he never asked the right questions and sought to get me off the stand relatively quickly. I guess he figured if he probed me too deeply, I'd make *him* look like a fool.

The plaintiff produced a video of his dead dog's triumph at a field trial. An expert on German Shorthaired Pointers testified that the live puppies would sell for at least two hundred dollars each (thirty years ago that was a lot of money). My testimony about the necropsy and the Montgomery County coroner's report were persuasive to the jury.

They awarded my client five thousand dollars for the loss of his dog.

***

One of my clients in Dayton, Ohio, had her own private zoo on a twenty-five-acre farm. She had sixty-eight cats, twenty cockatoos, parrots, monkeys, dogs, a cougar, and a llama. She had a reputation as an animal lover, and people would give her exotic pets that they could no longer tolerate. One day she called to tell me she was on her way to Kentucky to pick up a lion. She wanted to bring it directly to me to declaw and to remove the fangs.

I set up an appointment for Thursday afternoon.

She arrived with a bizarre story and a badly frightened lion. When she arrived in Kentucky and asked to see the lion, the owners led her to the kitchen closet. It seems that the lion was cute and cuddly when it was a little cub but had started to become aggressive as it grew larger. The owners became frightened, so they kept it in the kitchen closet and tossed food into the closet where the lion lived full-time.

As expected, after several months, the lion went crazy living in the closet with no fresh air and no socialization. My client and the lion owners in Kentucky spent four hours trying to coax the lion from the closet with ropes, sticks, and brooms. Finally, they succeeded in pushing and dragging the big cat into a cage. Then my client drove immediately to my office and presented the lion to me.

The lion was not happy to see me.

I was not happy to see him.

The big cat looked to be about a year old and capable of killing everyone in my animal hospital. However, we squeezed him into the back of his cage, and, through the wire of the cage, I was able to inject an anesthetic (ketamine) into the rear leg of the lion. Within minutes the big cat relaxed, and we lifted him onto the surgery table. There I put him on a gas anesthetic machine and removed his claws. It wasn't a whole lot different than declawing a smaller cat.

But the teeth (fangs) were another matter.

As this was my first dentistry on a lion, I asked the local dentist to assist. We were both surprised to learn that the fangs were hollow! By cutting off the tooth, we easily extracted the fangs by putting the points of the extractors inside and outside the cut tooth. The frightened lion recovered uneventfully but never quieted down and never became a good pet.

Despite my client's efforts to provide love and comfort, the lion did not recover psychologically from his months of living in the closet.

Owners of exotic pets always made me feel uneasy. I generally felt sorry for the pet and felt uncomfortable with an owner who seemed to be saying, "Hey, look at me, I own a lion (or cougar, or ocelot, etc.). Aren't I special?"

An exotic dancer in Dayton, Ohio, used a boa constrictor in her act. I never saw the act, but I'm sure it was tastefully done. One day at home, the dancer bathed the snake and was drying it with a large terry cloth towel when the phone rang. When she returned from the phone call, the snake had swallowed the towel. The snake recovered from surgery and returned to the stage.

Some years later I related this story to a veterinarian from India. He told me that, as a veterinary student in India, he had been presented with a boa constrictor that was impacted after swallowing a deer.

***

I was driving a U-Haul truck from Albuquerque, New Mexico, to Amarillo, Texas, and to limit my boredom, I picked up a hitchhiker in Tucumcari, NM. The hitchhiker crawled into my rented truck carrying a large canvas bag. The man had many tattoos and glass decorations in his nose and eyebrows. I learned that he was a tattoo artist and a drywall installer. He learned I was a veterinarian and immediately asked if I would examine his boa constrictor. "What boa constrictor?" I asked. He then surprised me by removing his wheezing snake from the canvas bag.

While driving the truck down the interstate highway, I examined the snake and recommended oral human interferon. I drove to my Amarillo home to acquire human interferon for the snake.

My wife was quite upset with me because the look of my hitchhiker was frightening to her. I treated the snake and took my hitchhiker to the bus station and bought him a ticket to complete his journey to Arkansas.

I advised him to keep the snake in the bag and not show it to any bus employees or bus passengers.

***

I always say animals are smart, often more intelligent than humans, and usually have greater street smarts than any politician. I observed a heroic cat that saw an emergency, then formulated and executed a flawless rescue plan.

At the age of fourteen, Crystal (my stepdaughter) had severe edema (swelling) in her ankles that was diagnosed as a sign of lupus erythematosus.

One night, Crystal's cat (who always slept with her) went into the bedroom of Crystal's mother (Ella Rose) and woke her by pawing at her face. Since that cat always slept with Crystal, Ella Rose immediately knew something was wrong and rushed to her daughter's room. Crystal was having convulsions. Ella Rose quickly called an ambulance, and Crystal was rushed to the hospital.

Tests revealed lupus had moved into her brain, causing cerebritis, thus accounting for the seizures. Crystal had lupus in her kidneys, joints, skin, and brain. Months of chemotherapy stopped the disease. Twenty-five years later, Crystal lives a full life, but with some residual kidney damage.

Crystal remains a cat lover and always keeps a pet cat. I think of that story often. This cat (Molly) recognized an emergency, then formulated a plan and executed it. Smart cat. Molly did not stop for a snack or make a litter box visit. Molly woke up Ella Rose, instead of me. Smart cat.

CHAPTER 3

The Russians Are Smart

I consider myself a patriotic American, but I'm also a man of science, which means I'm honor-bound to tell the truth. The Russians may not have beaten us to the moon, but they certainly beat us to the use of low-dose interferon. However, for reasons that remain unclear to me, this great discovery was abandoned, despite strong evidence of its effectiveness.

In 1969, Dr. V. Soloviev reported that approximately fourteen thousand people participated in controlled studies of placebo versus human interferon treatment during a natural outbreak of Hong Kong influenza. This is what he reported in a bulletin for the World Health Organization:

> The results are reported of experimental investigations and epidemiological observations on the use of human leukocyte-produced interferon for influenza prophylaxis. Field trials with interferon were carried out during the outbreaks of Hong Kong influenza in the USSR in January and February 1969. These covered about 14,000 people in comparable interferon and placebo-treated groups and achieved an effectiveness of between 56.3% and 69.2% according to the age group studied.[12]

Human interferon (about 128 units) or placebo was dripped into the nose daily for five days starting about the time of the first reported influenza cases. Human interferon treatment significantly ($P<0.01$) reduced the number of influenza cases.

Efficacy of leukocyte human interferon against Hong Kong influenza[13]

Group	Treatment	Number of patients		% sick
		Enrolled	Sick	
Adults	Interferon	2994	231	7.7
	Placebo	3129	551	17.6
Children 7–12 years	Interferon	1917	119	6.2
	Placebo	2055	413	20.1
Children 2–6 years	Interferon	463	22	4.8
	Placebo	454	53	11.7

In his discussion, Dr. Soloviev stated that "there are sufficient grounds to recommend human leukocyte-produced interferon as one of the means of influenza prophylaxis. The method is absolutely harmless, simple and convenient, and should be applied where there is an immediate threat of infection, that is, as a means of emergency prophylaxis."[14]

Dr. Soloviev reported that human interferon treatment was free of adverse events and proposed that human interferon "will be given proper place in the arsenal of means for fighting virus infections."[15]

In September 1971, a group of US scientists visited the Soviet Union and reported that there was advanced clinical work on the use of exogenous human interferon in Russia. This is what they reported in the *Journal of Infectious Diseases* in 1972:

> Although the visit was too brief to obtain an in-depth view of the multifaceted antiviral research program in progress, the visiting scientists obtained several lasting impressions. One of the most unexpected aspects of the Russian work is the advanced stage of development of interferon for clinical use. The group was surprised to find that human interferon prepared from leukocytes is already available for distribution through pharmacies in the Moscow area for use as a nasal spray against influenza. This exogenous human interferon has also been used to treat hepatic keratitis, varicella, adenovirus, and enterovirus infections. Although exogenous interferon is stated to be effective against these diseases, it is recommended for use only in high-risk groups against influenza.[16]

Did this Ohio farm boy ever think he'd be praising the Soviet Union for their health care? No, I did not. But you could walk into a Moscow pharmacy in 1972 and buy a nasal interferon treatment that had demonstrated positive effects against influenza. And if it worked against influenza, it also likely worked against a lot of other viruses.

Another group of US scientists arrived in Moscow on January 20, 1973, toward the end of an extensive influenza epidemic.[17] During the peak of the epidemic, January 8, the number of influenza cases reported in Moscow reached ninety thousand per day.

When treatment (500 IU leukocyte human interferon) was given by nasal spray three times daily for three days and then once daily for two days and started immediately after the first case of influenza, approximately a 60-percent decrease in influenza symptoms was reported in human interferon-treated patients.[18]

To achieve effects, human interferon was given by aerosol and orally. At the first sign of influenza illness, 600 IU human interferon was given over five minutes by the oral and nasal routes. This was repeated in two hours if the patient's symptoms were severe and was always followed by intranasal administration of human interferon twice daily for three days at the dosage used for prophylaxis. Clinicians reported that the human interferon treatment caused symptoms to disappear quickly; fever and headache cleared immediately.

Merigan et al. reported in *The Lancet* in March 1973 on the treatment of eleven human volunteers with placebo and eleven volunteers with human interferon (eight hundred thousand units given by spray-gun with a nozzle yielding a fine mist).[19] The human interferon was given in divided doses twenty-four, five, three, and one hour before challenge with 10,000 TCID50 of influenza B/Hannover/1/70 virus.

Human interferon therapy did not alter the subsequent frequency or severity of infection as judged by clinical signs, symptoms, sero-conversion, and frequency or intensity of virus shedding on days three and four after infection.

The dose and schedule used by Merigan were different from those used in the Soviet Union. The dose of human interferon given to eleven volunteers in one day by Merigan was approximately one thousand times higher than the dose given to thousands of subjects in the Soviet Union. The prestige of Merigan and his coauthors, the high quality of Merigan's study, and the prevailing belief that "more is better" cast doubt on the Soviet results.

Treatment of eleven flu patients by Merigan of Stanford University was considered by most readers as more meaningful than thousands of flu cases treated in Russia.

Too bad.

The world missed out and would continue to suffer.

In 1976, Dr. Veselina Arnaoudova reported on "160 units" of human interferon given five times a day for three days (therapeutic) or 160 units given three times a day for three days repeated twice at ten-day intervals (prophylactic), for acute respiratory viral infections. From the abstract:

> Children with respiratory virus infections and healthy contacts were given leukocytic interferon (LIF) as treatment and prophylaxis. The results emphasize the effectiveness of the treatment when LIF was given on the 1st and 2nd day after onset of illness. In these cases, the course was mild and of short duration and without complications. In children who did not receive LIF, the clinical course was more severe, the disease lasted longer, and complications were not uncommon. Prophylactic instillation of LIF succeeded in preventing the disease in 85.2% of the contacts; those who contracted the infection had a mild disease with insignificant complications. No allergic or intoxication phenomena were observed in children receiving LIF.[20]

Allergic or adverse events were not observed in the 868 children, including newborns and premature babies, given human interferon during a natural outbreak of influenza. It was reported that human interferon therapy reduced the severity and duration of disease. It was reported that oral human interferon was effective in preventing influenza.

In their review of clinical trials in 1979 with exogenous human interferon, Dunnick and Galasso reported on Soviet research by stating that "Human interferon is said to prevent or ameliorate influenza A when a relatively small amount is administered intranasally as a spray over a period of three to four days."[21] The authors then pointed out that other studies needed much greater amounts of human interferon (even 14 billion IU) to show an antiviral effect against other respiratory viruses. Still others reported that rapid muco-ciliary clearance mechanisms would predict the need for much larger doses of human interferon.

The clinical observations from Russia and Bulgaria did not fit the "more-is-better" view of human interferon therapy.

Imanishi et al. reported in Japan in 1980 that intranasal drops of human interferon (5,000 IU/daily) for four months reduced the frequency and severity of disease due to influenza A and parainfluenza virus:

> Drops of a human leukocyte-derived human interferon preparation were instilled into the nasal cavity of 86 volunteers once a day from the beginning of December 1977 to the end of March 1978. The rise in hemagglutination-inhibition antibody titers against the three strains of influenza viruses was not significantly different between the interferon and the control group.
>
> However, the rise in antibody titers against the parainfluenza virus was less in the interferon group than in the control group. There were fewer complaints of fever resulting from upper respiratory tract disease in the interferon group. Moreover, fevers higher than 39 degrees C were found only in placebo recipients. Fourteen of the 41 volunteers in the interferon group complained of subjective symptoms due to upper respiratory tract infection, whereas 28 of the 43 volunteers in the placebo group complained. This difference was significant.
>
> Thus, our study indicates that prophylactic administration of an human interferon preparation can influence upper respiratory tract disease.[22]

Data were collected on eighty-three volunteers in the study. Fever occurred in six of forty volunteers given human interferon and in fifteen of forty-three volunteers given placebo ($P<0.01$). Symptoms such as headache, cough, fatigue, anorexia, myalgia, etc., occurred in 34 percent of volunteers given human interferon and in 67 percent of volunteers given placebo ($P<0.01$).

In 1982, Isomura et al. reported that human interferon (10,000 IU/day) or placebo was dripped into the nostrils of twenty-seven children daily for sixty days. The children lived in an orphanage where natural outbreaks of influenza occurred during the treatment period. This is what they reported:

> A double-blind, controlled trial to ascertain the preventive effect of human interferon-alpha on upper respiratory viral infections was performed on children in a closed community. Drops of human interferon were instilled into the nasal cavity of 13 healthy children aged one to three years. Fourteen children were given placebos as controls. Administration of the interferon and clinical observations were carried out in the winter of 1980.

Serological examination revealed that this was the period of outbreaks
of influenza type A epidemics in the community. Clinical manifestations
referable to influenza virus infection were milder in the interferon-treated
group than in the controls. However, there was no significant difference in
the serological responses of the two groups after infection with influenza
virus type A.[23]

Human interferon at this greater dose did not prevent illness but significantly reduced the duration of fever and the mean peak fever. Clinical
manifestations of influenza were milder in children given human interferon.
Adverse events were not observed.

During influenza epidemics in 1983, 1984, and 1985, Jia-Xiong et al.
treated 140 children with a spray of natural human interferon into the nose
and mouth twice daily for three to four days. The total daily dose was
reported to be 700-1600 IU.

The fifty-three control children were given traditional Chinese herbs.
Children given human interferon had significantly (P<0.01) faster nor-
malization of temperature after the first treatment. Clinicians reported
that pharyngitis and lymphadenosis of the pharynx improved when fever
subsided.

In 1987, Dr. Jia-Xiong and fellow researchers published their observations about interferon alpha, based on their seven years of research:

> In 1979, the Pediatric Department of Shanghai Changzheng Hospital, in
> collaboration with the SINE Laboratory, succeeded for the first time in pro-
> ducing human leukocyte interferon aerosol (IFN-*a* aerosol). The biological
> value of the interferon used in the aerosol was not adversely affected. Over
> 80% of its particles have diameters under five um. In animal experiments no
> irritative reactions were observed. Clinical trials showed the preparation was
> effective in children's respiratory viral disease including influenza, bronchi-
> olitis caused by RSV, mumps, asthma (infectious), asthmatic bronchitis and
> recurrent upper respiratory tract infections.
>
> Based on our animal experiments and clinical trials lasting more than
> 7 years, IFN-*a* aerosol is proved effective, safe, and non-traumatic. It has no
> side effects and is economical and easy to use.[24]

I need to spend a little time talking about interferon aerosol and why
that type of nasal spray might be a highly effective delivery system. We
believe interferon is produced in the nasal passages because that is likely to

be the first line of the immune defense system. Since we are likely to inhale pathogens, it makes sense that Mother Nature would put a guardian at that gate to our respiratory system. (Pathogens we might eat in food are likely to be broken down by our digestive juices.) While I believe interferon taken by mouth is effective, there is good reason to believe a nasal spray could be as effective. The intranasal route was not tried in cats, dogs, or pigs because animals tend to vigorously object to anything placed up their nose.

In 1985, Saito et al. reported in Japan that leukocyte (50,000 IU human interferon/day) was sprayed into the nasal passages of thirty-seven human volunteers twice daily for eight consecutive weeks. Placebo was given to thirty-six volunteers in this double-blinded, controlled study. Dosage compliance was a problem such that only eleven out of seventy-three volunteers administered over 90 percent of their medication.

After excluding volunteers who took less than 50 percent of their medication, those volunteers who developed the common cold, and those who had a rise in antibody titers before the study, it was concluded that the rise in complement fixation antibody titers against influenza A was not significantly different between treatments:

> Human leukocyte interferon (IFN-a), 50,000 IU per day, was sprayed into the nasal cavity of 73 volunteers twice a day from January 9 till March 4, 1984. The rise in complement fixation antibody titers against influenza A virus was not significantly different between the interferon group and the placebo group. However, the number of subjects without elevated antibody titers and without symptoms in the interferon group was significantly higher than that in the placebo group ($p < 0.05$). Prophylactic nasal spray of IFN-a seems to protect against upper respiratory viral infections.[25]

What this research showed is that both groups in the course of their daily lives were exposed to influenza, both developed antibodies, but those who received interferon had lower antibody titers and no symptoms, meaning their immune system was easily dealing with the virus. Again, by following nature, we are figuring out how to deliver health to people.

In 1984, Dr. Phillpotts et al. in the United States reported that thirteen human volunteers self-administered lymphoblastoid human interferon (2.7 million IU/dose) three times/day for four and a third days, starting one day before a challenge with influenza virus A/Eng/40/83. Illness occurred in four of thirteen volunteers given human interferon and in ten of seventeen volunteers given placebo (not significant).

Serological responses and/or virus recovery were obtained in eleven of thirteen volunteers given human interferon and in fourteen of seventeen volunteers given placebo (not significant). Mean daily nasal secretion weights and mean clinical score were lower in the human interferon group but were significantly different from the placebo group only on postchallenge day two. In their discussion section, the authors wrote:

> This study provides clear evidence that intranasally administered lympo-
> blastoid IFN effectively prevents the illness of experimental rhinovirus in
> human subjects. The severity of clinical symptoms and virus replication were
> reduced by IFN treatment, although a proportion of volunteers was infected
> and showed a serological response.[26]

In 1987, Treanor et al. reported on intranasal delivery of human interferon to sixteen volunteers experimentally challenged with influenza A/California/78.[27] The dose of human interferon was 5 million IU twice daily delivered to each nostril by a hand-held metered pump spray. The spray was self-administered starting forty-eight hours before virus challenge and continued for seven days. A total of 70 million IU human interferon was given to each human interferon-treated volunteer. Nine volunteers were given placebo.

Treanor et al. reported that illness developed in five of nine placebo recipients and in three of sixteen human interferon recipients (P=0.087). There was no significant difference in median total score between treatment groups, although scores in human interferon recipients were consistently lower. None of the volunteers developed a fever.[28]

Eight of nine placebo recipients excreted influenza virus compared to thirteen of sixteen human interferon recipients.

Placebo recipients excreted virus on 64 percent of test days compared to excretion of virus on 41 percent of test days by human interferon recipients. Treanor et al. concluded that intranasally administered human interferon had a "definite antiviral effect" and a "possible clinical effect" in the prophylaxis of experimentally induced influenza A virus infection.

Hayden et al. reported that a daily intranasal spray of 8.4 million IU of human interferon for twenty-eight days, and a total study dose of 235 million IU of interferon, was too toxic and "not feasible for prophylaxis of respiratory virus infection."[29] Hayden's study enrolled twenty-six volunteers to receive human interferon and twenty-four volunteers to receive placebo.

Phillpotts et al. reported that eleven volunteers were given thirteen equal doses of 2.7 million IU human interferon (3 times daily for 4 1/3 days) but *not* challenged with virus. Mild "nasal symptoms" developed in five of eleven human interferon-treated volunteers, compared to one of eleven volunteers given placebo.[30]

Greenberg and Harman presented three possible explanations for the large in vivo dose requirements of intranasally applied human interferon.[31]

First, the mucus overlying the nasal epithelial cells could contain a substance or substances that inactivate or inhibit the applied human interferon before it could render the cells resistant to virus challenge.

Second, the mucus barrier or rapid clearance mechanisms could prevent the human interferon from reaching the nasal epithelial cells.

Third, the nasal epithelial cells lining the nasal cavity could be insensitive to human interferon compared with tissue culture cells.

A series of experiments that addressed each of these possibilities was published and made the use of low-dose human interferon theoretically unacceptable for influenza.

The skepticism about the Soviet clinical research with human interferon in influenza is summarized by Cantell in his 1998 book, *The Story of Interferon*.[32] On page 220, Cantell wrote:

> My attitude towards these results, like that of probably all other interferon workers in the western world, was extremely skeptical from the very beginning. Over the course of years, many visitors brought me interferon which they had purchased in the Soviet Union: we found that this was not only very impure but also contained only a few hundred units of interferon per milliliter.
>
> Our crude interferon preparations contained at least a hundred times more, and our concentrated interferon at least a hundred thousand times more interferon than those in the Soviet pharmacies. When glasnost and perestroika finally led to the break-up of the Soviet Union, these interferon preparations silently vanished from the shelves of Russian pharmacies. I do not know whether they had ever done any actual harm to the recipients, but they certainly did not enhance the reputation of interferon, or the status of Soviet biomedical science.[33]

Cantell and others dismissed the observations made on thousands of influenza cases because they could not believe such trivial doses of impure interferon could have an effect. After all, Merigan et al. showed that eleven volunteers given 800,000 units of human interferon were not protected against a challenge of influenza B.

Phillpotts et al. failed to demonstrate a benefit in thirteen volunteers given 35 million units of human interferon and then challenged with influenza A.

Treanor et al. could barely demonstrate a benefit from 70 million units of human interferon given to sixteen volunteers before they were challenged with influenza A.

The observations by Soviet scientists made no sense to Western scientists when compared to studies in which high dose human interferon failed to provide a clinical benefit.

Did anyone in the West test human interferon administered by low dose?

No. If they did so, they did not publish the results.

Testing was not conducted using low doses of human interferon because "theoretically," it would not work, and did not fit the wobbly "more interferon is better" philosophy.

When Japanese scientists tested low dose human interferon, benefits were generally reported. Imanishi et al. reported a benefit from 5,000 IU human interferon daily.[34] Isomura et al. reported clinical benefits from 10,000 IU human interferon daily.[35] However, when Saito gave volunteers 50,000 IU human interferon daily, significant benefits were not seen.[36]

The Soviet and Bulgarian claims of a benefit from a few hundred units, Jia-Xiang's claim of a benefit from 700-1600 units, Imanishi's claim of a benefit from 5,000 units, and Isomura's claim of a benefit from 10,000 units human interferon should be reconsidered.

Low-dose oral or intranasal human interferon should be tested rigorously using the pure human interferon formulations available today. Influenza is too important to dismiss all the low-dose human interferon data.

Our oral HBL human interferon influenza study was conducted in two hundred volunteers in Australia. The paper was published in the September 2013 issue of *Influenza and Other Respiratory Diseases*.[37] It was titled "Low-dose oral interferon alpha as prophylaxis against viral respiratory illness: a double-blind, parallel controlled trial during an influenza pandemic year." Enrollees took placebo or 150 IU of HBL human interferon each morning.

Post hoc analysis of groups noted significant reductions in the incidence of acute respiratory illness in males, those aged fifty years or more and those who received seasonal influenza vaccine. The overall incidence of acute respiratory illness was not limited, but HBL human interferon reduced the severity of signs and symptoms.

Millions of people have gotten sick with influenza and died since Dr. Soloviev reported his important observations more than fifty years ago. Had experts embraced his research instead of denigrating it, much suffering might have been avoided.

In my opinion, Dr. Soloviev deserved consideration for the Nobel Prize for his work, instead of the scorn he received.

CHAPTER 4

Involvement with Interferon

My job in Cincinnati as a veterinarian lasted less than one year, as I yearned to return to college in 1970, using the GI Bill to fill in what I considered serious "gaps" in my knowledge. I was accepted into graduate school at the University of Missouri, College of Veterinary Medicine, Department of Veterinary Microbiology.

At that time, Dr. Ray Loan was department chairman and Dr. Bruce Rosenquist was my graduate program advisor. These two men are among the best I have ever met in veterinary medicine. I'll always be grateful for their kindness as I stumbled toward proficiency. They were patient, wise, and fair in their treatment of all their graduate students.

It was at the University of Missouri in 1970 that my career in research and my understanding of interferon began. The earliest research with cattle interferon had been conducted by Dr. Rosenquist, and I became his first graduate student, conducting research on the bovine interferon response in the blood of calves given IBR (infectious bovine rhinotracheitis) virus.

After a year and a half, I tired of the rigors of graduate course work and research and decided to start my own veterinary practice in the Dayton, Ohio, area. Animal hospitals and kennels were then usually zoned into industrial areas of cities and competed with service stations and convenience stores for zoned space.

This difficulty created an opportunity that I turned to my advantage.

In three years, I opened three veterinary hospitals in succession and sold them to other veterinarians who did not want to fight the zoning laws and other obstacles inherent in building a veterinary hospital. I found I enjoyed the challenge of creating something from scratch more than the daily grind of practice.

After pocketing the money from the sale of the veterinary hospitals and knowing I could always make money in the field, I returned to the University of Missouri to complete my graduate work.

Upon returning to Missouri in 1975, I began studying the interferon response of cattle in the nasal secretions, instead of the blood. I'd routinely spray IBR virus into the nose of susceptible calves, then collect their nasal secretions by stuffing a tampon up their nose. After ten to fifteen minutes, I'd fish out the string end of the tampon and pull out the tampon.

The tampons were often quite juicy, and I'd place them into a large syringe. The syringe was then placed into a vise equipped with a hole from which the nasal secretions would flow after pressure was applied on the syringe plunger. Various methods were employed to collect nasal secretions including vise grips, pliers, and a lime press. But the vise and syringe were the easiest to operate and the most efficient method to extract the contents of the tampon.

There's no doubt in my mind, I was the greatest and most talented collector of "cow snot" in the state of Missouri, if not the United States of America west of the Mississippi River.

Some might say that was a dubious honor.

After graduating from the University of Missouri in 1978, I was hired as an assistant professor to teach veterinary virology at the University of Illinois (UI). I arrived in the fall, carrying with me the nasal secretions containing interferon collected during my three years of research at the University of Missouri. I taught veterinary virology to sophomore veterinary students for two years at UI and tried to develop an independent interferon research program.

***

On Christmas Eve 1976, my mother-in-law showed me her ankle where a small mole had enlarged into a black cauliflower-shaped mass about one inch in diameter. I told her that if I saw such a mass on a dog, I would remove it. After the holidays, a diagnosis of malignant melanoma was made, and surgery to remove the melanoma and regional lymph nodes was performed in March 1977.

From April 1977 to April 1978, she received chemotherapy and laetrile, and she also modified her diet. But by October 1978, she again had tumors palpable on her leg.

In 1978, it was known that the interferon from cattle and humans was quite similar. One publication from the journal *Science* in 1972 titled "Interferon Administration Orally: Protection of Neonatal Mice from Lethal Virus Challenge" drove my thinking that interferon might help my mother-in-law with her difficulties:

> Interferon was identified in the milk of mice injected with an interferon inducer. The kinetics of interferon appearance in serum and milk were similar, but maximum concentrations in milk were 10 to 20 percent of those in serum. Interferon administered orally to neonatal mice were detected in their serums. Significantly more newborns survived an oral challenge with vesicular stomatitis when interferon had been induced in the lactating mothers.[38]

The earlier dogma that interferon was species specific was undergoing reevaluation and was eventually rejected. During my PhD research, hundreds of samples of nasal secretions had been collected from calves, and the samples were assayed for interferon activity.

Because my mother-in-law was frightened and desperate about her cancer, I offered her the bovine interferon from my nasal secretions collection in October 1978.

Before I gave the bovine interferon to my mother-in-law, I swallowed a few milliliters to assure her that the material was safe (it tasted salty because it had been dialyzed in phosphate-buffered saline).

A wart (3 mm diameter) had been on my left index finger for more than ten years. After ingestion of nasal secretion interferon, the wart began to feel and look different within three days. It had a red ring around it. Within three weeks, the wart (caused by human papillomavirus) completely regressed and has not returned.

It has since been published (primarily from Italy) that low-dose oral human interferon is a good treatment for papillomavirus, specifically articles by Montevecchi,[39] Palomba and Melis,[40] Bastinaelli,[41] Biamonti,[42] and Verardi.[43] No need for that HPV vaccine!

My mother-in-law had to be desperate or crazy to take semipurified bovine nasal mucus (cow snot), containing bovine interferon activity. Because the nasal secretions samples were not pure enough to inject, I recommended she swallow the samples three times a day and made up a schedule of five days of treatment followed by a week of abstinence.

Over the course of about two months, she swallowed all the nasal secretion interferon I had collected over three years. She ingested 3–10 ml three times daily over four five-day trial periods. The schedule was designed to roughly mimic the swallowing of nasal secretions by cattle during viral respiratory tract disease.

By February 1979, all but one of her tumors had regressed, and the largest and last tumor disappeared by June 1979.

She was free of malignant melanoma and lived for another twenty-five years.

Although intriguing, the improvement in my mother-in-law and my own wart regression could have been unrelated to bovine interferon treatment. As the scientific community would no doubt say, the results were "anecdotal," but I've always liked the response that "the plural of anecdote is data."

I was impressed by the regression of my wart and my mother-in-law's recovery, but my colleagues dismissed both observations as coincidental. And a part of me didn't want to be too open about what I'd done. Maybe it would simply be my "secret weapon" to be deployed if anything threatened my close friends or my family.

But I certainly didn't want to put my career on the line.

However, in the fall of 1979, I attended a conference sponsored by the New York Academy of Science where the "antiviral aspects" of interferon were presented and the mechanisms by which interferon stimulated various components of the immune system were discussed. I left the conference fired with enthusiasm to try oral bovine interferon in another malignant melanoma patient.

Interferon was no longer just an antiviral protein.

Now I was aware that interferon was also an immune modulator.

I contacted my mother-in-law's osteopathic physician in Ohio and asked if he had more melanoma victims desperate enough to swallow cow snot. The physician replied that he had two such patients. Calves were purchased and inoculated intranasally with IBR virus.

Their nasal secretions were collected, dialyzed, and centrifuged and then assayed for interferon activity. I collected the nasal secretions from cattle in Illinois just as I had previously collected those Missouri nasal secretion samples prepared for my mother-in-law and supplied the samples to the physician.

One melanoma cancer patient died before the interferon was ready, but the other patient began the treatment with cow snot in February 1980.

The remaining malignant melanoma patient had tumors covering more than 20 percent of her lower leg. The tumors had been surgically removed twice. However, after each surgery, the tumors recurred. The overweight, elderly woman with hypertension had refused amputation and was being treated with laetrile of unknown benefit. The tumors were increasing in size and were oozing, which necessitated wrapping of the leg in diapers. The oozing was so copious she had to change the diapers several times during the day to avoid leaving puddles of fluid on the floor.

Within a week of initiating the oral bovine interferon treatment, the leg dried noticeably, skin coloration improved, and the "burning" pain in the leg subsided. Over the next couple of weeks, the largest tumors lost volume and flattened.

Over the course of more than two years, her tumors regressed.

During her therapy, the bovine nasal secretion interferon was used for the first two-week period because that was all I had collected. I then began to use my laboratory to produce interferon in bovine fetal kidney (BFK) cells using bluetongue virus (BTV) as an interferon inducer. The purpose was to produce in the laboratory the same kind of interferon that calves produced in their nasal secretions but using a method that was easier and less expensive. At that time, I mistakenly assumed that interferon in the nasal secretions was the beta type, when in fact it is predominantly an alpha type. The predominant type of interferon produced by BFK cells induced by BTV is interferon type beta. However, when bovine white blood cells are induced by BTV, interferon alpha and other interferon types are also produced.

Based on these two patients treated with bovine nasal secretion interferon, Drs. DasGupta and Schade at the University of Illinois Medical Center in Chicago agreed to cooperate in a pilot human clinical trial in June 1980. In August 1980, I moved to Amarillo, Texas, to work for Texas A&M University (TAMU). I spent my first year in Texas producing bovine interferon beta in the laboratory for the cancer patients in Chicago. The research was supported by Johnson & Johnson (J&J) in the USA and Kureha Chemical Industry Co. Ltd. of Tokyo.

The BFK cells induced by BTV produced mostly interferon beta, which was supplied to Dr. DasGupta, instead of interferon alpha from nasal secretions. Five terminally ill cancer patients were given bovine interferon beta but failed to show improvement, so the tests were discontinued.

This was the end of my first attempt to treat human melanomas with oral interferon.

I lost the financial support and the interest of J&J and Kureha because of the failure of oral interferon beta to improve melanoma patients.

The University of Illinois patent representatives also lost hope and did not find any other interested companies.

When these five melanoma patients given laboratory-produced interferon beta failed to respond like the earlier two patients on interferon from nasal secretions, I conducted two simple tests and realized that my laboratory-produced interferon was quite different from the interferon found in nasal secretions. I heated each kind of interferon and noted that the one from BFK cells was destroyed by heat (56°C for thirty minutes) and the other was not.

I also placed both interferons on cat tongue cells in cell culture and noted antiviral activity for interferon from nasal secretions, but not the interferon from BFK cells. After a year's worth of work, I realized I had been wasting everyone's time and had treated the melanoma patients with the wrong interferon.

By the time my mistake was realized, my credibility with the clinical investigators was destroyed, and melanoma research ceased and was not resurrected until 2017, when bovine interferon alpha was again somewhat efficacious in a melanoma woman in San Francisco, California.

***

In 1980, ten other human patients tried oral interferon beta (BFK cell origin) for warts, moles, etc., and observations were made that the interferon beta was well tolerated and did not seem to produce any adverse effects. Needing more cases and a more uniform disease condition to treat, I started to treat feline leukemia virus-associated diseases. Many cats with feline leukemia virus were readily available and were treated with interferon given orally.

The use of natural bovine interferon beta in cats with naturally occurring cases of feline leukemia virus-associated diseases began on July 2, 1980. Seven cats were treated by Dr. Mary Tompkins of the University of Illinois and Dr. Barbara Stein of the Chicago Cat Clinic. Three cats died when given steroids preceding or during interferon therapy. Those four cats given only interferon beta (without steroids) underwent remarkable clinical improvements, as reported in 1982 in the journal *Feline Practice*:

> There is no known agent at present that will reverse FeLV-associated non-regenerative anemia. In this report four cases of NRA in FeLV-positive cats recovered from the anemia after treatment with supernatant fluid from the bluetongue virus-infected bovine kidney cells containing bovine beta interferon. In each case, the cat began to eat and be more active within the first week of treatment. The CBC's began to improve after the second treatment in all cases and have remained normal for up to 17 months post treatment.[44]

One of the concepts I've become familiar with over the years is that the body has an amazing capacity to heal itself, if you can address the problem.

There's no possible way I can make the claim that interferon cures anemia. What I can say is that the feline leukemia virus can cause anemia, so that when the virus is under control, the cat's own body may fix the anemia. I question whether we fully understand the role of long-term persisting viruses in many of our most challenging health conditions.

***

Because of the unique clinical observations, an invention disclosure was made to the University of Illinois. A patent application was filed in 1980 titled "Delivery of Biologically Active Components of Heterologous Species Interferon Isolates" and eventually issued as US Patent No. 4,462,985. By the time of that first patent application, bovine interferon alpha and interferon beta were consumed by twelve people, eight mice, five cats, four guinea pigs, three hamsters, and a dog.

Plans were made to continue trials in people and cats and to start a clinical trial on pigs. There had been no evidence of adverse events attributed to oral administration of bovine interferon alpha or interferon beta, and I was optimistic that an important discovery had been made.

Some cats suffering from feline leukemia showed clinical improvement after interferon consumption, and two people with malignant melanoma had reported clinical improvement.[45] Changes had occurred in some warts of people consuming bovine interferon alpha. One woman with acne reported normal skin after oral interferon treatment. Another woman reported that a painful lump on her toe became painless and reduced in size after interferon alpha consumption.

After moving to the research center in Amarillo, which was part of TAMU, I asked an Aggie engineer to build a hydraulic tampon squeezer.

I'd envisioned a small device, but the Aggie engineer had a grander vision and delivered a hydraulic device that was over three feet tall, weighed over a hundred pounds, and was double-barreled (it held two syringes at once). The device generated more than 300 psi (pounds per square inch) of pressure and was extremely efficient at extracting nasal secretions from tampons. It's amazing in science how often a good engineering solution can lead to breakthroughs.

Now because of one smart Aggie engineer, I was truly the world's leader in "cow snot" collection. Nobel Prize in Science, here I come!

Because the double-barreled Aggie tampon squeezer was too bulky and heavy to transport, I used a small vice from a hardware store for tampon squeezing away from the lab. During one experiment, my colleague, Dave Hutcheson, and I traveled to Newport, Tennessee, to purchase calves. We wanted to ascertain how much interferon the calves had in their nasal secretions before shipment to Texas, so we bought ten boxes of regular-size tampons and collected nasal secretions in Tennessee. The tampons were placed in 50-ml syringes and taken to our motel room for extraction of the nasal secretions.

After dinner, we returned to our room, kicked off our shoes, and began squeezing out the tampons with the portable vice. We had more than a hundred tampons and made a big mess in the motel room. We had syringes, tampons, and vials scattered everywhere and had a large cooler to transport our supplies home.

We were about half-finished with the extraction of the nasal secretions when a car alarm went off in a car parked in front of our motel room. We looked outside and could not tell why the alarm was activated. The police soon arrived and knocked on our door to ask if it was our car.

We must have been an unusual sight to this Newport policeman as he stood in the doorway of our room. Syringes, tampons, and vials were on the floor, bed, and cabinets.

The policeman surveyed the scene and was briefly quiet. For a moment he must have wondered what type of drug operation he'd stumbled into but must have decided we looked far too geeky to be doing anything illegal.

Strange but not criminal.

He asked if the car belonged to us, we told him no, and he left. I've always wondered if such scenes were common to this policeman in East Tennessee, or if he dismissed us as eccentric Texas Aggies, and it made for a good story to tell his family at dinner that night.

***

In 1970, I started conducting research on interferon. By 1979, I was determined to learn as much as possible about the phenomenon of oral human interferon and assumed it would be easy to demonstrate the benefits to skeptics.

I was wrong. It has not been easy. It's been incredibly hard, and I don't understand why.

Even now, the notion that oral human interferon has biological activity is dismissed by most scientists. Unfortunately, most scientists, when thinking briefly about the oral administration of human interferon, assume it is not possible for a protein to be orally active because it would be digested.

Generally, clinicians and scientists believe that human interferon can't be given orally and that it can't work in low doses. However, my research confirms the early work of the Russian Soloviev (see beginning Chapter 3) whose important research in influenza has been largely (and wrongly, in my opinion) dismissed.[46]

To summarize the world's literature on oral interferon in one document that could be shared with interested scientists, a special issue of the *Journal of Interferon & Cytokine Research* (*JICR*) was edited by Manfred Beilharz and published in August 1999. That issue was devoted to the subject of oral administration of interferon and a few other cytokines. There were twenty-one articles on various aspects of the research, and three of them were mine.

It was greeted with a big yawn by the research and medical establishment.

During six and a half years at the Texas A&M Agricultural Experiment Station (TAES) in Amarillo, Dave Hutcheson and I conducted fifty-one controlled studies involving 2,857 calves. Thirty-seven of the fifty-one cattle studies dealt specifically with the production or use of interferon alpha in cattle. At the end of 1987, I left Texas A&M to devote myself full-time to the commercialization of oral interferon. The results of those cattle studies formed the early basis of my understanding of how best to use interferon orally.

Dose titration studies of human interferon in calves have consistently predicted that 0.1-1.0 IU/kg of body weight is the best dose of human interferon for the prophylactic treatment of IBR virus. The mechanism by which such a seemingly trivial amount of human interferon (in comparison to massive doses injected into humans) given by the oral route is beneficial to calves is not completely understood.

But articles in the August 1999 issue of the *Journal of Cytokine Research and Interferon Research*, devoted mostly to interferon, should give you an understanding of how many different conditions might be treatable with interferon. I have included a sampling of those articles so you may more fully appreciate the potential wide reach of this therapy.

One article was titled "Immunomodulation and Therapeutic Effects of the Oral Use of Interferon-*a*,"[47] and others were titled "Autoimmunity Is a Type I Interferon Delivery-Deficiency Syndrome Corrected by Ingested Type I IFN Via the GALT System,"[48] "Oral Use of Interferon-*a* Delays the Onset of Insulin-Dependent Diabetes Mellitus in Nonobese Diabetes Mice,"[49] "Suppression of Late Asthmatic Response by Low-Dose Oral Administration of Interferon-*b* in the Guinea Pig Model of Asthma,"[50] and "Low-Dose Oral Use of Human Interferon-*a* in Cancer Patients."[51]

The foreword by Dr. Philip I. Marcus, a professor in the Department of Molecular and Cell Biology at the University of Connecticut, brilliantly laid out the challenge to those of us working with interferon:

> The earliest reports of the use of orally administered [interferon (IFN)] languished because they were received by a skeptical scientific community. Scientists had reasoned that IFN was too susceptible to enzymatic destruction to exert a significant biological effect following oral administration. Certainly, it would not be active at the low doses reported to be effective. This skepticism was bolstered by an apparent lack of species specificity for IFN administered by the oral-pharyngeal route and the absence of a biological/molecular basis for the effects reported for this route of administration.
>
> The IFNs were discovered because of their antiviral effects. Soon, it became clear that these effects were manifest by the action of only a few molecules per cell. Thus, the potential was there from the very beginning to have IFN do what is now reported by the proponents of orally administered IFN, namely at low doses under physiological conditions . . . To this end, this Special Topics Issue is dedicated to the pioneers of the oral use of IFN and is a tribute to their tenacity in staying the course.[52]

Perhaps because interferon alpha subspecies from different animals are so similar (but may bind somewhat differently to cell receptor sites), and because nasal secretions are naturally swallowed, human interferon or bovine interferon can be used orally in the treatment or prevention of some diseases. Controlled data in animals convinced us to study oral human interferon in humans.

It's not as if we're advocating a toxic drug.

On the contrary, we recommend human interferon at a dose and route of administration near to that which man and animals swallow in their nasal secretions during respiratory tract infections.

Conditions in which human interferon is produced in millions or billions of IU probably do not exist in nature. When human interferon is given to man or animals in high doses, it is invariably toxic.

A search of the Internet will reveal comments by many patients who claim their lives have been ruined by high-dose injected human interferon. Our three-radiolabel studies indicate interferon receptors are located in the oral and pharyngeal mucosa and, I believe, support our use of human interferon orally.

Before traditional scientists will believe that human interferon (as it is used by some veterinarians) is effective in any condition of man, human controlled studies must be conducted. Ultimately, the treatment costs would be low and could significantly benefit an AIDS or cancer program even if it does no more than stimulate appetite (as it does in animals).

***

Since cats given human interferon for their feline leukemia virus-associated disease seemed to experience an increased appetite, I set out to specifically test oral human interferon as an appetite stimulant in cattle on October 25, 1980. Additional research on appetite and feed efficiency and the use of oral human interferon resulted in two patent applications at TAMU that eventually issued as US Patent Nos. 4,497,785 and 4,820,515.

Clearly, human interferon given orally in low doses stimulates the appetite, as demonstrated by the successful registering of our two patents.

A surprising example of appetite stimulation occurred in a Tennessee calf. From an order buyer in Tennessee, a hundred calves were purchased for a study. Some of the calves were given human interferon orally before shipment to Texas to assess the effect on respiratory tract infection. One of the calves was already sick with respiratory tract disease when he was treated in Tennessee. Upon arrival in Texas, the calf stumbled off the truck and fell. The calf was dehydrated, depressed, and had severe respiratory disease.

With some prodding, the calf got up and walked into the chute, where he was given a Pinpointer® ear tag, which allowed monitoring of his feed intake. The calf then walked to the feeding trough, ate about five and a half pounds of feed, and dropped dead. Up until that time, I would never have believed a calf so near death would eat.

Apparently, oral human interferon stimulated the appetite of the dying calf.

***

I pestered the patent attorney who was handling the issued University of Illinois patent (US Patent No. 4,462,985) to find a pharmaceutical partner to develop oral human interferon. Eventually, the patent attorney consented to license the patent to me if I would write a business plan, which I had no clear understanding on how to accomplish. I figured we'd form a company, give 10 percent ownership of the company to the University of Illinois, and pay a 4 percent royalty. I eventually got the job done, and we formed a corporation called Amarillo Cell Culture Company, Inc. (ACC) in 1984.

One of the veterinarians, Dennis Moore, cooperating in the successful treatment of feline leukemia and canine parvovirus, made the initial investment to create the company. Dennis served on the ACC Board of Directors for thirty years and was very supportive of our research because he had used low dose oral human interferon safely and successfully in his own California veterinary practice.

Dennis had a favorite cat given to him by an uncle. The cat had feline leukemia, and Dennis tried every treatment, without success. Dennis was persistent and tracked me down regarding low-dose human interferon. The cat made a fast recovery when human interferon was provided. Dennis said he wanted to make a donation to my favorite charity. After I told him my favorite charity was low-dose oral human interferon, he said he would invest.

Without initial support from Dennis Moore, oral Human interferon research in man would not have occurred in the United States. In 1984, Dennis and I started ACC. In 1996, before our initial public offering (IPO), the name of the company was changed to Amarillo Biosciences, Inc. (ABI).

Dennis was a farm boy who attended Colorado State Uninversity to obtain his degree in veterinary medicine, and I was a farm boy who attended Ohio State to obtain a veterinary degree. We met because of feline leukemia and founded a biotech company that raised more than forty million dollars and conducted research in twenty countries with human interferon.

Where did we raise that money? Our Japanese partners, some sales, license fees, private placements, grants, our IPO, and loans kept ABI going until bankruptcy. Billy Walters of Whale Securities managed the IPO. He introduced us to the famed physician Dr. Claus Martin of Germany, who personally invested over a million dollars. Billy Walters and his family also invested. Many people reached into their pockets to try and help ABI. But when the government and pharmaceutical companies don't seem like they want you to reach the finish line, even a large amount of money like forty million dollars eventually runs out.

***

A young Food and Drug Administration (FDA) investigator came to my Amarillo office and announced her intentions to inspect my facility that was marketing an oral human interferon approved by the Texas Department of Health. I phoned my FDA consultant, a former FDA employee who had been head of compliance for veterinary medicine. Even though I was an intrastate operation, my consultant advised me to allow her to inspect my laboratory. She spent four hours visiting my small facility (only about 500 sq. ft.) and then called her boss in Dallas. I heard her tell her boss over the phone that she could not find anything wrong. After discussing the situation with her boss and obtaining some instructions, she asked to see my records on salt purchase.

After looking at my salt purchase invoices, she announced, "I think I can *get* you." She then asserted the salt used in my product had crossed state lines. Even though the human interferon was manufactured in Texas, the bottle was manufactured in Amarillo, the label was printed in Amarillo, and sales were only to Texas veterinarians, the FDA threatened to take me to court unless I ceased all Pet Interferon Alpha sales. They claimed that had to happen because the salt, purchased in Texas, had originally been mined outside of Texas. The woman said that there are no salt mines in Texas (not true). I "voluntarily" ceased sales of Pet Interferon Alpha rather than undergo the expense of a battle with the FDA. I wondered why the FDA bothered to stop the sale of a product doing so much good in pets.

An animal health company in the United States worked with me on a specific cat disease in 1991–1992. Another animal health company with headquarters in France worked with me in 1995. Both companies conducted excellent research in animals but did not find a significant economic benefit due to oral human interferon.

The French company noted a significant ($P < 0.05$) appetite enhancement in healthy pigs, as previously reported in our issued US Patent No.

4,497,795. Even though a benefit was noted, the benefit was not good enough to hold the interest of the French company.

In the past twenty years, much has been learned, and I can look back with regret at the higher doses of bovine and human interferon often tested. Our 2016 publication titled "Low-Dose Oral Interferon Modulates Expression of Inflammatory and Autoimmune Genes in Cattle" in *Veterinary Immunology and Immunopathology* taught us the importance of using the correct low dose of oral interferon. It was a remarkable study, using the latest technological tools to tease out the exact effect low dose natural bovine interferon was having on gene expression vital to fighting viruses. My fellow researchers included scientists from Ohio State University, Texas A&M University, and a private laboratory in Rhode Island. This is how we described what we were investigating:

> In the cattle study reported herein, the expression of 92 immune response-related genes was analyzed. The mode of action of oral interferon on gene expression in peripheral blood mononuclear cells (PBMCs) was evaluated further by assessment of interferon effects on families of genes in the Kyoto Encyclopedia of Genes and Genomes (KEGG) cytokine-cytokine receptor interaction pathway.[53]

There are probably some things to explain. Because of the advance of technology in mapping the human genome and determining what each one of those genes do in the body, we have a pretty good idea of which genes are responsible for guiding the response of the immune system. It's the immune system that determines how effectively we can combat viruses and other pathogens.

And to take matters even further, it seems that most of the damage comes from the body's response to these pathogens, the storm of inflammatory molecules known as cytokines.

In the conclusion section of the article, we recounted the findings, especially how low-dose natural bovine interferon modulated the expression of twelve genes within a specific pathway linked to cytokine expression:

> Collectively, extensive *in vivo* data have demonstrated that orally administered low-dose interferon has rapid and systematic beneficial biological effects in animals and humans. In the present study, it was demonstrated that oral interferon statistically modulated the expression of 12 genes within a specific KEGG pathway (cytokine-cytokine receptor interactions). In turn, these data

suggested that low-dose oral interferon exerts systematic and largely beneficial response genes and their products likely needed for recovery from any
viral disease, including foot and mouth disease virus.[54]

It seems dramatic to me how a medication that was developed in the
1980s and subjected to a nullification campaign by the medical establishment is still showing such dramatic results nearly forty years later with the
latest technology.

Science supports the use of interferon to modulate the immune system
to appropriately respond to viruses and not damage the body through a
cytokine storm. In the COVID-19 crisis, we are understanding how damaging the cytokine storm can be in causing damage. Why are we not looking at interferon in our current crisis?

CHAPTER 5

Interferon Gets Hot

Studies in cattle, performed under various conditions, demonstrated that oral human interferon is safe and has antimicrobial effects, as well as modification of expression of genes and effects on metabolism (such as weight gain) in cattle. Unlike the high doses injected into humans, the optimal oral dose in cattle for these positive effects is usually about 1 IU/kg body weight of oral human interferon once per day. Many studies reporting the benefits of oral human interferon in cattle have been published over the years. Representative studies are briefly summarized below.

Over seven thousand feed lot calves weighing 182–295 kg were enrolled into a placebo-controlled study to assess the effect of a *single* oral dose of natural human interferon at about 0.7 IU/kg body weight. The human interferon, along with standard antibiotic therapy, was given when cattle were first diagnosed. This is the abstract for that research:

A total of 4577 feeder cattle were treated with a single oral dose of 33.0 international units (IU) of natural human interferon alpha (Human interferon alpha) per 100 pounds of body weight, upon entry into the feedlot hospital pen. Another 2494 cattle received diluent alone and served as placebo controls. Cattle were evaluated on the number of treatment days required until the animals could be returned to the feedlot, and the death loss of cattle in the hospital pens. The mean number of days of treatment for control animals (3.9 days) was greater than in cattle treated with human interferon alpha (3.1

days). Mortality of placebo treated animals was 5.9% versus 3.6% (P < 0.001). These data suggest that oral low dose human interferon alpha is of benefit in reduction of feedlot-associated morbidity and mortality.[55]

When we look at the death rate of control cattle at 5.9 percent and the cattle treated with a single oral dose of interferon at 3.6 percent, that means the interferon cattle had an almost 40 percent drop in mortality. The sample size was more than ample to tease out the effects, especially as it is confirmed by multiple other studies. How great is that to a rancher's bottom line?

The *single* oral dose of human interferon treatment reduced death in feedlot cattle by nearly 40 percent. That's a very dramatic result. Should this therapy be tested in humans with respiratory disease? The cost of the human interferon was less than a dime. Is that why low dose oral human interferon is not acceptable in human medicine? It is too cheap and effective?

Two hundred and sixty-four Holstein bull veal calves weighing 40–50 kg were enrolled in a placebo-controlled study. Recombinant human interferon (500 IU/calf) was added to their milk replacer once daily for five consecutive days:

> Low doses of recombinant human interferon alfa 2a (Human interferon alpha 2a), were given orally in milk replacer formula to veal calves to determine the efficacy of [Human interferon] alpha 2a for protection against diarrhea, ear and/or respiratory tract infections common in vearling operations. Calves given [Human interferon] alpha 2a had fewer days and a lower incidence of diarrhea, compared to placebo-treated calves. Calves treated with [Human interferon] alpha 2a had significantly (P < 0.05) fewer ear infections and fewer total days of ear infection than did placebo-treated calves.
>
> The mortality rate was lower in the [Human interferon] alpha 2a treatment group (1.6%) than in the placebo treatment group (2.9%) and calves given [human interferon] alpha 2a had a greater average weight gain (13.1 lbs. more per calf) than calves given placebo.
>
> These data demonstrate that orally administered Human interferon alpha 2a exhibited a protective effect against clinically significant signs of disease in veal calves, reduced the mortality rate in this population, and enhanced average weight gain.[56]

The human interferon treatment significantly protected against diarrhea and otitis media (inflammatory diseases of the middle ear) during the following fifteen weeks. Calves given human interferon had fewer diarrhea days and reduced severity of diarrhea, compared to placebo-treated calves.

Moreover, oral human interferon treatment reduced the incidence of otitis media from 18.7 percent in placebo-treated controls to 8.9 percent. The mortality was lower (1.6 percent, compared to 2.9 percent), which translates into a 45 percent drop in expected mortality. Additionally, the average weight gain/calf was 6 kg (a little more than thirteen pounds) better in human interferon-treated calves.

Is it becoming clear how much natural human interferon is adding to the health and survival of young cattle? The reasonable question is whether young humans would experience similar benefits.

With Dr. Hutcheson and the Texas A&M University Agricultural Research Station, we performed our own research on eighty-four IBR sero-negative light weight (182–227 kg) feeder steers, which were enrolled in a dose titration study of placebo versus 40, 200, or 400 IU of human interferon/calf in the treatment of IBR virus infection.[57] Calves were given the human interferon treatment at the time of challenge with IBR virus, and half the calves additionally were given human interferon treatment a day before IBR virus challenge. An oral dose of 40 IU human interferon, but not higher doses of natural human interferon, resulted in significantly less fever than in calves given placebo.

Dr. Hutcheson and I also conducted research on two hundred steers and bulls (mean weight 209.4 kg) that were purchased from six sale barns in the southeast USA.[58] The calves were treated for three consecutive days in Tennessee with placebo or 0.11, 1.1, or 11 IU of human interferon per kg of body weight before shipment to an experimental feed lot in Bushland, Texas. Oral doses of natural human interferon at 1.1 IU, but not higher doses, per kg of body weight had a significant beneficial effect on lowering fever. Calves with normal temperatures treated in Tennessee with doses of 0.11 or 1.1 IU/kg body weight (but not 11 IU) had a significant weight gain twenty-one days after arrival.

Natural human interferon was tested orally in twenty-four one-year-old Friesian bulls challenged with virulent Theileria parva, a protozoan.[59] In two experiments, natural human interferon orally at 1 IU/kg body weight protected eleven of twelve bulls against fatal theileriosis compared to the death of five of eight placebo-treated.

Two natural human interferon alpha preparations, (human interferon alpha [Cantell]) and (human interferon alpha [ISI]), were used for the oral

experiment of cattle experimentally infected with Theileria parva. In the first
experiment, 8 Friesian bulls were inoculated with a 1 in 10 dilution of a sporo-
zite stabilate of T. p. parva (Marikebuni) stock. Four of the cattle were treated
daily with 1 international unit/kg of body weight (i.u./kg bwt) of human
interferon alpha (Cantell) from day -2 to day 8 p.i. None of the 4 calves given
interferon developed clinical theileriosis, but 3 of the 4 control cattle died of
theileriosis while the fourth control had a mild infection.[60]

Let's put that in simple terms we can all understand. In the first experiment, 100 percent of the cattle treated with interferon survived exposure
to this parasite, while only 25 percent of the cattle who did not receive the
treatment survived. When we increased the numbers, we found a similar
pattern. More than 90 percent of the interferon treated cattle survived,
while only 37.5 percent of the noninterferon-treated cattle survived.

Low dose orally administered natural human interferon (from
Hayashibara Biochemical Labs [HBL]) at 0.5 IU/kg body weight was given
in Japan to treat rotavirus diarrhea in calves.[61] The researchers reported that
human interferon-treated calves had significantly less severe diarrhea, shorter
duration of diarrhea, better weight gain, and reduced rotavirus excretion.

Using cDNA microarrays, a study of eight matched Japanese Black
calves identified genes differentially regulated in bovine peripheral blood
after oral therapy with 1.0 IU/kg of human interferon.[62]

Perhaps this points to why it should not be surprising that appetite, feed
efficiency, and metabolic changes have been reported in cattle after low-dose
oral human interferon administration.

In addition to research using oral human interferon as an appetite stimulant in cattle and as an immune modulator in shipping fever, oral human
interferon was used to treat cats with feline leukemia and dogs with parvovirus diarrhea ("parvo"), as I will discuss in the next section. Cats with feline
leukemia and dogs with parvo were so common and numerous that veterinarians were willing to cooperate in treating cases encountered in their
practice.

***

In 1986, an excellent study involving researchers from the University of
Illinois and The Ohio State University used twenty-one eight-week-old specific pathogen-free (SPF) cats that were hysterectomy-derived from a breed-
ing colony maintained at The Ohio State University and looked at the use of

human interferon.[63] The pathogen-free cats were innoculated intravenously with the Rickard strain (subgroup A) of feline leukemia virus (FeLV). Natural human interferon from Immuno Modulator Labs (IML-Stafford, TX) was given orally once daily to fourteen cats, and seven control cats were untreated. These were our findings:

> Low doses (0.5 or 5.0 U) of human interferon alpha human interferon given orally prevented the experimental development of fatal feline leukemia virus (FeLV)-related disease. Twenty-one FeLV susceptible cats were inoculated with the Rickard strain of FeLV. Cats given oral human interferon survived significantly (p <0.001) longer than untreated FeLV-infected cats. Moreover, only 4 of 13 (30.8%) human interferon-treated cats developed clinical disease during the course of the study, whereas 100% of the untreated control cats developed fatal FeLV-related disease. Thus, in experimental retroviral disease, heterologous species human interferon provided significant clinical benefits.[64]

It is time to discuss feline leukemia virus, its connection to human leukemia in predominantly Asian populations, and the connection to HIV-AIDS. I also want to examine what it might mean if current controversial research on another family of animal retroviruses from mice is confirmed. These are referred to as XMRVs or xenotropic murine leukemia viruses, and some research has indicated that they may have jumped the species barrier into human beings.

While science had long shown that retroviruses could cause disease in animals, it wasn't until 1980 that scientists confirmed a retrovirus could cause disease in humans. That retrovirus was called HTLV-1 (human T-cell leukemia virus) and was discovered in samples from Japan by a team including Robert Gallo, Frank Ruscetti, and Bernie Poiesz. A retrovirus is made of RNA rather than DNA and uses the enzyme reverse transcriptase to translate its RNA into DNA so it can integrate into an organism's genetic code.

HTLV-1 may have originally been FeLV but somehow jumped into human beings in Japan during the nineteenth century. During that time, it was common for the Japanese to eat cats, and it's believed by some that's how transmission occurred. Only about 5 percent of humans who carried the virus, though, would go on to develop leukemia. In most cases, the person's immune system could suppress the damaging effects of the virus.

The HIV retrovirus (HIV-1) was identified in 1984 by a team led by French scientist Luc Montagnier, for which he won the Nobel Prize in 2008. That was a monkey virus, probably from a chimpanzee, which had somehow jumped the species barrier into human beings. For that retrovirus there were few individuals spared its damaging effects.

The XMRV retrovirus was first identified in prostate cancer tissue samples in March 2006, by a team led by Robert Silverman of the Cleveland Clinic. It seems clear that the precursor to this human infection came from a mouse, by a method that remains unclear. Later research linking this retrovirus to chronic fatigue syndrome/myalgic encephalomyelitis, autism, and other conditions is still ongoing and very controversial.

So, as you can see, FeLV aided the investigation into the HIV-AIDS epidemic and may yet provide insights into XMRV, or any other retroviral threat to humanity that may develop in the future. Looking at the effect of interferon in feline leukemia virus in cats may provide clues as to whether interferon would be an effective treatment for retroviruses in humans.

Okay, so now that I've spent some time discussing interferon usage for a cat leukemia virus, let's review how good the results were from that experiment.

The human interferon was given to nine cats at 0.5 IU and to five cats at 5 IU once daily for seven consecutive days every other seven days for six months. All seven control cats died on the average of seventy-three days after the FeLV inoculation. Six of nine cats given 0.5 IU of human interferon lived at least 550 days after FeLV inoculation. Three of five cats given 5.0 IU oral human interferon lived significantly longer than control cats.

Interestingly, these surviving cats lived for years even though the oral human interferon was given for only six months. These surviving cats were carriers of FeLV but were asymptomatic. In other words, they were living with the virus, rather than dying from it.

In an oral human interferon study conducted in Europe years later, thirty adult cats naturally infected with feline immunodeficiency virus were enrolled to oral treatment of natural human interferon (twenty-four cats) or placebo (six cats).[65] The human interferon was given at 10 IU/kg of body weight once daily every other seven days for six months.

Two months later, treatment was repeated for months eight through fourteen. Only one of twenty-four cats given oral human interferon died, and only one of six control cats lived. Not only was there a significant survival benefit, but the human interferon-treated cats experienced a rapid improvement in their clinical condition and improvement in some blood variables:

Feline immunodeficiency virus sustains an AIDS-like syndrome in cats, which is considered a relevant model for human AIDS. Under precise enrollment requirements 30 naturally infected cats showing overt disease were included in a trial of low-dose, oral human interferon-alpha treatment. Twenty-four of them received 10 IU/Kg of human interferon-alpha and 6 placebo only on a daily basis under physician supervision. The low-dose human interferon-alpha treatment significantly prolonged the survival of virus-infected cats (p <0.01) and brought to a rapid improvement of disease conditions in the infected hosts . . .

As shown in other models of low-dose interferon-alpha treatment, there was a rapid regression of overt immunopathological conditions in virus-infected cats. This hints at a major role of interferon-alpha in the control circuits of inflammatory cytokines, which was probably the very foundation of the improved clinical score and survival despite the unabated persistence of virus and virus-infected cells.[66]

These data suggest a major role for human interferon in the control circuits of inflammatory cytokines, which likely explains the improved clinical score and survival despite the cats still having the virus. The benefit to cats occurred when human interferon was given orally at low doses (10 IU/kg or less) compared to millions of IU of human interferon injected into humans. Again, we have the perplexing thinking of injecting interferon at doses hundreds to thousands of times higher than those naturally found in nature.

Cats given natural human interferon or interleukin-2 (IL-2) orally at Virginia Tech had significantly increased red blood cell counts. The human interferon treatment was more effective than IL-2 in terms of duration of beneficial effects. Human interferon was given at doses of 0.05, 0.5, or 5.0 IU/cat (6 cats per group) on days zero, three, seven, ten, fourteen, seventeen, twenty-one, and twenty-eight. Significant increases in red blood counts, hemoglobin, and hematocrits occurred at days fourteen, twenty-one, and twenty-eight in cats given human interferon at 0.5 or 5.0 IU/cat. Monocyte counts were also significantly increased on these same days, but only in cats given 0.5 IU of human interferon:

These preliminary results indicate that very low doses of human IL-2 and human interferon administered orally, induce significant changes in various hematological values of cats, mainly those associated with red blood cells. The effects by both cytokines were dose dependent. Human interferon was more effective, both in terms of dose and duration of treatment. The number of monocytes were elevated significantly only by one, the middle, dose of human interferon indicating a selective optimum for this effect. The biological importance or use of these findings remains to be determined.[57]

In a clinical study that was not published, but included in a patent applications nineteen practicing veterinarians made diagnoses of feline leukemia virus-associated diseases and treated cats with either bovine interferon beta or human interferon. Cats were treated orally for five consecutive days, two to three times daily. After five to seven days without treatment, cats were treated for another five days. This five-day treatment was repeated three to four times, in most cases.

Sixty-nine cases of feline leukemia virus-associated diseases, free of lymphosarcoma, were diagnosed over four years. Cats given human interferon had a significantly higher survival rate at six and twelve months after treatment than cats given bovine interferon beta.

In my opinion, both published research and clinical practice have demonstrated that human interferon is of enormous benefit in many conditions among animals. I think the positive results among cats suffering from the effects of FeLV demonstrate proof of concept that interferon can probably have strong effects among humans suffering from retroviruses like HIV, or possibly even the newly discovered XMRVs.

***

Three practicing veterinarians made diagnoses of sixty-eight cases of naturally occurring canine parvovirus based on clinical signs and laboratory data. The vets used either human interferon or bovine interferon beta orally three to four times daily for two to three days in the treatment of the dogs.

Only 69 percent of forty-two dogs given bovine interferon beta survived, compared to the survival of 92 percent of twenty-six dogs given human interferon, a significant benefit for human interferon in the treatment of canine parvovirus.

Five litters of newborn puppies were purchased in 1984–1985 to test the efficacy of low-dose oral human interferon during canine herpesvirus (CHV) disease. Not much benefit was noted; puppies died quickly when infected with canine herpesvirus. Oral human interferon is probably not useful in every disease and was not beneficial in these test conditions. Newborn puppies have a large surface area and a small body weight, so they are easily chilled and can't adequately maintain their body temperature when they vomit or have diarrhea. Perhaps better temperature control would have yielded a better result. Moreover, the dose of human interferon might have been too large for their relatively small body weight and immature immune system. Other diseases caused by herpes viruses in mice, cats, cattle, and humans respond favorably to treatment with low-dose oral human interferon.

Working with small animal practitioners prior to 1986, we observed the beneficial use of low-dose oral human interferon in treating dogs and cats. During the late 80s, we tested human interferon in many different conditions in dogs and cats, and we compiled a list of fifteen animal diseases and conditions that seemed to respond to oral human interferon. Pet Interferon Alpha was low-dose human interferon approved by the Texas Department of Health for use by Texas veterinarians in the treatment of feline leukemia and canine parvovirus.

Pet Interferon Alpha was the first approved interferon for pets in the world. In addition to its approved uses for feline leukemia and canine parvo, veterinarians using Pet Interferon Alpha orally and topically reported clinical improvement in:

1. Feline Respiratory Tract Disease
2. Feline Infectious Peritonitis
3. ADR (Ain't Doing Right)
4. Appetite Depression
5. Lupus Erythematosus
6. Canine Herpesvirus
7. Feline Distemper
8. Canine Distemper
9. Kennel Cough
10. Rodent Ulcer
11. Canine Warts
12. Collie Nose
13. Arthritis
14. Pyoderma
15. Demodex mange

A significant survival benefit was noted in dogs infected with canine parvovirus treated with oral human interferon, compared to bovine interferon beta. The human interferon or bovine interferon beta was sprayed in the mouth at least three times daily even if the dog was vomiting.

The effectiveness of oral natural human interferon in the treatment of naturally occurring, immune-mediated canine keratoconjunctivitis sicca (KCS) has been evaluated in one study. Dogs with KCS were treated orally with human interferon once daily by owners as the sole therapy. The dogs were examined every two weeks for the duration of the trial of twelve weeks.

Each dog was given escalating doses (20, 40, 80 IU) of human interferon. A favorable response was observed in 55 percent (eleven of twenty) of all treated dogs. Clinical findings in responding dogs included increased wetting of the eyes, decreased mucus discharge, and fewer signs of discomfort. Seven of eleven dogs with favorable outcomes had an increased Schirmer's tear test (STT) of at least 5 mm/min after treatment with oral human interferon. As a group, the KCS-treated dogs had a posttreatment STT (10.5± SEM 1.4 mm/min) significantly greater than pretreatment baseline values. All dogs that responded did so with the 20 or 40 IU, not 80 IU, dose of human interferon. Side effects were not reported, and all dogs tolerated the treatment:

> In summary, it appears that low-dose orally administered human interferon induced a positive beneficial effect on both the subjective and objective manifestations of canine KCS. Interferon therapy, unlike cyclosporin A, is without side effects, is easy to administer, and could be a significant economic advantage for treating canine KCS. It will be important to determine if oral use of human interferon ameliorates the chronic manifestation of canine KCS with long-term therapy.[68]

KCS in dogs is like Sjogren's Syndrome in people, and later tests of human interferon in patients with Sjogren's Syndrome in Japan and the United States determined that oral low doses of human interferon were safe and efficacious.[69]

***

In a prominent Canadian veterinary journal in 2004, it was reported that natural (50 IU) or recombinant (90 IU) human interferon given orally once daily for five days significantly reduced relapses in horses with inflammatory airway disease (IAD).[70] A total of thirty-four horses were in the study, divided up into three roughly equal groups. While many horses will naturally recover from IAD, there was a pronounced difference between the human interferon-treated group and the control group. From the discussion section of the article:

> Despite the initial response rates, over the following two weeks as horses returned to normal activity, there was a recurrence of clinical signs in some horses. The rate of relapse or return to clinical signs in the subsequent two weeks was significantly higher in the placebo-treated group (55%) than in the interferon-treated group (less than 10% in both groups).
>
> Only those horses that remained free of clinical signs on day 28 were considered to have been cured. Overall, approximately 80% in each of the interferon-treated groups compared with 35% in the placebo group were considered to have been cured. The significant difference in relapse rates supports the hypothesis of a significant beneficial clinical effect of interferon alpha in IAD [inflammatory airway disease].[71]

Let's talk about some of the differences between the various groups. The rate of relapse in the control group was five times higher than in the human interferon-treated group. Additionally, 80 percent of the human interferon group cleared the infection, while only 35 percent of the control group was cured of disease.

Although the study contained a relatively small number of horses, the results were so encouraging that they certainly should have been followed up with trials with larger numbers. From 1996 to 1997, investigators from The Ohio State University studied the use of low-dose human interferon in twenty-two horses with IAD, one of the most common afflictions among racehorses, causing impaired performance, interruption of training, and premature retirement. We found a significant positive effect and suggested mechanisms of action:

> Oral administration of low-dose (50 U) human interferon lowered total nucleated cell counts in BALF [bronchoalveolar fluid] from horses with IAD [inflammatory airway disease] for 15 days. Because the etiopathogenesis of IAD has not been identified, the mechanism of therapeutic benefit of [human interferon] is unknown. However, human interferon is suspected to reduce pulmonary inflammation via immunomodulation or antiviral activity. Human interferon induces an antiviral state in target host cells by stimulating production of enzymes that inhibit antiviral protein synthesis and degrade viral DNA (Weigent et al., 1984; Stanton et al., 1987).[72]

Using the same group of twenty-two horses, there was a study on whether there was an accompanying decrease in their inflammatory markers as a result of treatment with human interferon. From the summary of the paper that was published in a different veterinary journal in 1997:

> Total protein, IgG and IgA concentrations in BALF were reduced (P<0.05) 8 days after administration of 50 and 150 IU human interferon and 15 days after administration of 50 IU of human interferon. Procoagulant activity after albumin concentrations in BALF were lower 8 days after administration of 50 IU human interferon. Oral administration of low dose human interferon appeared to ameliorate these parameters of lower respiratory tract inflammation in Standardbred racehorses with IAD.[73]

Almost every time low-dose oral human interferon is investigated, it seems there is a significant positive benefit associated with its use. It's important to state again that the point of this information is to urge pharmaceutical companies to follow nature and determine what leads to health, rather than using science to create some super-charged, high-priced wonder drug.

Low-dose oral human interferon has been shown to be a natural and effective way to treat many inflammatory diseases. The use of low-dose oral human interferon is consistently safe and efficacious in different animal diseases.

***

Oral natural human interferon at 36 IU per piglet given treatment daily from birth to three days of age significantly reduced mortality of piglets from endemic Streptococcus suis infection at Berea College, Kentucky (Dr. Carolyn Orr, not published). Moreover, oral natural to human interferon at 10 or 20 IU/day for four days significantly reduced transmissible gastroenteritis virus (TGE) mortality in young piglets in Pennsylvania in another study in which I participated:

> During a natural outbreak of transmissible gastroenteritis (TGE), groups of piglets were treated orally for 4 consecutive days with placebo or 1.0, 10.0 or 20.0 international units (IU) natural human interferon alpha (nHuIFNα). Piglets that were 1–12 days of age and given 1.0, 10.0 or 20.0 IU nHuIFNα had significantly ($P < 0.01$) greater survival rates than placebo-treated piglets;

survival rates were the greater for the highest level of nHuIFNα treatment. In contrast, beneficial effects of nHuIFNα were not observed in piglets farrowed during the disease outbreak and given nHuIFNα within hours of birth. Oral nHuIFNα therapy modulates the natural course of high morbidity and mortality commonly seen with TGE.[74]

Newborn piglets given 1 IU of human interferon had better survival rates from the infection than did placebo-treated piglets, but the survival benefit was not statistically significant. TGE mortality is caused by the ensuing diarrhea, and this study of 1,740 piglets supports the use of oral human interferon in COVID-19. TGE virus is a coronavirus.

Since low-dose oral human interferon can save some pigs from TGE where death and diarrhea are so pronounced, low-dose oral human interferon probably should be tested for COVID-19, as it seems that another complication is the ensuing respiratory problems. Here is what I argued in a 1999 review article on interferon:

> From its modest beginnings 20 years ago, the interest in the uses of the oral route for interferon and other biologically active signaling molecules has grown dramatically. The currently accepted method of interferon administration by injection of millions of IU is obviously not physiologic; it does not duplicate the levels of this potent biological regulatory factor as found in vivo. Indeed, the notion that low doses of oral administered interferon may mimic natural disease processes occurring during respiratory infections has theoretical appeal.[75]

Yes, I'm making the argument again that it's low-dose interferon that has the best chance of producing a significant effect because it models what happens in a healthy organism. If we can get a virus under control, we shift the immunological battle in the body toward health, rather than sickness. The respiratory complications will be less severe if the levels of the virus in the body are lower.

Significant modulation of rotavirus infection was seen in infected piglets given natural human interferon in North Carolina. A positive dose-dependent correlation was noted between the human interferon dose and weight gain in weanling piglets. Colostrum-deprived piglets given 50 IU/kg body weight of human interferon had less rotavirus four days after virus inoculation. This is from the abstract of our study:

Colostrum-deprived neonate piglets challenged with rotavirus and 3-week-old newly weaned piglets naturally exposed to rotavirus were treated with low doses of natural human interferon alpha (nHuIFN alpha) administered into the oral cavity or included in the liquid diet. The colostrum-deprived piglets given the highest dosage of nHuIFN alpha (50 IU/kg body weight) had lower viral excretion scores at 3 (p less than 0.11) and 4 days (p less than 0.001) after virus inoculation.[76]

Time after time we are getting the same result. Interferon primes the body to respond in the way it should by marshaling or modulating the resources of the immune system in a rational manner. Young pigs have an immature immune system that is not fully developed to respond to the pathogens to which they are exposed. One might say interferon gives their system a temporary maturity, so it can more effectively respond.

In studies enrolling 1,448 pigs undergoing natural outbreaks of porcine reproductive and respiratory syndrome (PRRS) or post-weaning systemic wasting disease (PMWS), human interferon was given in the feed at 10 IU/kg body weight for ten consecutive days. The research in Italy by Amadori and colleagues is outstanding and should be embraced by the swine industry. This is what they reported:

The results obtained with this treatment can be summarized as follows:
- A dramatic improvement of both clinical score and growth rate in problem herds characterized by high losses at weaning (>10% dead and culled piglets).
- A significant improvement of the growth rate only in herds characterized by better welfare and management conditions.
- A considerable reduction of drug usage in most herds under treatment.

In conclusion, the results obtained in the above field trials confirm that a new approach to prophylaxis of PRRS and PMWS in piglets is badly needed; this must imply a proper combination of hygienic measures and immunological treatments aimed at restoring the full potential of homeostatic anti-inflammatory response.[77]

Let's really dig into this research. PRRS and PMWS can cause a loss of greater than 10 percent among pig populations. The oral use of interferon caused a "significant improvement" in their growth rates. And, if you're growing at a healthy rate, you're less likely to die. Simple logic, right? And if you're healthy, you're far less likely to need a drug intervention. Again, simple logic.

In pigs (unlike most other animals), the optimal oral dose of human interferon may be as high as 10 IU human interferon/kg body weight. Even so, the dose given by injection to humans is usually thirty thousand times higher than the oral dose to pigs. The human injection is usually 3,000,000 IU, so if the human patient weighs 220 pounds (100 kg), an injected dose of 3,000,000 IU would be 30,000 IU/kg (Physician's Desk Reference, 2020). One can clearly see how this is overkill of a bad idea.

CHAPTER 6

Interferon in Humans

In 1987, Hoffmann-La Roche reported that they had radiolabeled their human interferon and injected it into five cancer patients.[78] Some of the intravenously injected human interferon, surprising the authors, made its way into the mouth, nose, and paranasal sinuses of the patients.

I believe human interferon should be administered into the mouth or nose in the first place without requiring it to work its way from the injection site.

Low-dose orally administered human interferon has been reported to be safe and effective in treating diseases in dogs, cats, horses, cattle, pigs, poultry, mice, rats, and humans. Low-dose orally administered human interferon, without toxicity, has been reported to be helpful in managing cancer, viral diseases, and autoimmune diseases in humans.

In 1992–1993, a clinical trial demonstrated that low-dose oral human interferon was safe and effective in treating measles in children. The study was conducted in the Philippines using tablets made with lymphoblastoid human interferon from the UK company Burroughs Welcome. The study was published in the *Journal of Interferon and Cytokine Research* in 1998.

To determine the safety and effectiveness of low-dose oral human interferon against measles, thirty pediatric patients were randomly assigned to either placebo or human interferon treatment and observed daily for fourteen days. The patients in the human interferon group received a daily sublingual dose of 200 IU of lymphoblastoid human interferon. The human interferon-treated group had a shorter average duration of malaise (3.2 vs. 10.7 days, p < 0.0001), anorexia (3.1 vs. 6.7 days, p < 0.0001), and irritability

(1.1 vs. 2.2 days, p <0.01) and shorter duration of macular or maculopapular/ papular lesions (4.3 vs. 8.2 days, p <0.0001) and branny desquamation (4.6 vs. 5.8 days p >0.05) and shorter time for rash to become generalized (5.5 vs. 10.3 days, p <0.0001).

Let's break this study down. Measles is much more dangerous in Third World countries than in First World countries in which better nutrition and hygiene is generally thought to lessen the severity of the disease. The Philippines, where the study was conducted, has significant poverty, which likely means measles is more dangerous in the Philippines than in the United States. In that study:

1. The duration of malaise decreased from 10.7 days to 3.7 days.
2. The duration of anorexia decreased from 6.7 days to 3.1 days.
3. The duration of irritability decreased from 2.2 to 1.1 days.
4. The duration of macular/maculopapular/papular lesions decreased from 8.2 to 4.3 days.
5. The duration of desquamation (skin peeling) decreased from 5.8 to 4.6 days.
6. The duration in which a rash might go from being localized to one that is generalized went from 10.3 to 5.5 days.

In addition, there were no hematologic, renal, or liver toxicities observed. That means this was a highly successful intervention with no observed side effects. Do you know how rare that is in all of medicine? And yet, it is consistent with what has been observed in all the years of research with low-dose oral interferon.

Hematologic, renal, or liver toxicities were not noted. It therefore appears that low-dose oral lymphoblastoid human interferon used in this pilot study was both safe and effective in children with measles infection.[79]

This study and other publications on the safety and effectiveness of low-dose oral human interferon were ignored by pharmaceutical companies that were intent on developing high-dose injected human interferon.

It has been published that interferon is safe and effective in treating sometimes fatal coronavirus infections in pigs[80] and has been strongly recommended for use in essentially all mammals in a 2019 publication:

Type I interferons (IFN-I) generally protects mammalian hosts from virus infections, but in some cases, IFN-I is pathogenic. Because IFN-I is protective, it is commonly used to treat virus infections for which no specific approved drug or vaccine is available. The Middle East respiratory syndrome-coronavirus (MERS-CoV) is such an infection, yet little is known about the role of IFN-I in this setting. Here, we show that IFN-I signaling is protective during MERS-CoV infection. Blocking IFN-I signaling resulted in delayed virus clearance, enhanced neutrophil infiltration, an impaired MERS-CoV-specific T cell responses. Notably, IFN-I administration within 1 day after infection (before virus titers peak) protected mice from lethal infection, despite a decrease in IFN-stimulated gene (ISG) and inflammatory cytokine gene expression.[81]

This article is supportive in several different ways of the importance of interferon to fight viruses. Blocking interferon signaling renders an animal more susceptible to the damage caused by a virus. Enhancing interferon, by providing what is essentially a "booster" dose, protects against the inflammatory cytokine storm that is at the heart of how viruses cause their destruction.

This is a common biological defense mechanism that I believe can be used safely in all mammals, including humans. Many others would come to that same conclusion, as well.

***

The low doses of oral human interferon used in dogs and cats from 1980 to 1985 led to the filing of patent claims in November 1986 in what was eventually issued as US Patent No. 5,019,382. Other claims in that patent filing related to human claims that arose from observations I personally made or that Val Hutchinson, MD, of Amarillo made in cooperation with me.

Dr. Hutchinson became interested in the use of oral interferon in late 1982. He visited me because he had heard about the use of oral interferon in treating cats. He had a medical colleague "dying" of malignant lymphoma, and he wanted to give his friend oral interferon. I suggested a dose and timing of oral interferon administration, which Dr. Hutchinson followed in successfully treating his friend; the complete clinical success of that treatment and the patient's complete recovery are detailed in US Patent No. 5,019,382.

A friend of Dr. Hutchinson, Dr. Bill Hale, started using oral interferon in the treatment of Hale's aged mother who had mesothelioma. She lived fifty-nine months after starting oral interferon, an unusually long survival time for a case of mesothelioma. Dr. Hale reported that during times of oral interferon administration, his mother was more active and had less fluid buildup in her chest. Dr. Hale. also used low-dose oral interferon in his Ear, Nose, and Throat medical practice. He reported that cases of mononucleosis would not become chronic in patients given oral interferon.

With his initial success, Dr. Hutchinson became enthusiastic about the use of low-dose oral interferon and treated many human cases of rheumatoid arthritis, multiple sclerosis, common cold, and warts and, in 1990, even published his results regarding aphthous stomatitis, a condition characterized by benign and noncontagious mouth ulcers:

> Recombinant human interferon alfa-2 ((HuIFN alpha) was administered orally once daily in a low concentration (1,200 IU/day) to nine patients with chronic recurrent aphthous stomatitis (RAS), and a placebo solution was given to 10 control chronic RAS patients in a double-blind study. All HuIFN alpha-treated patients had total remission of their aphthae within a 2-week period, while placebo control patients had no change in their condition. The 10 placebo control patients were then treated with HuIFN alpha in a manner identical to that used for the initial principal group. Within a 2-week period, all original placebo patients had complete remission of their aphthae.[82]

Several things stand out to me about this study. The medical community agrees that mouth ulcers can be caused by viral infections, as well as other conditions such as Behçet's disease, vitamin B-12 deficiency, or dietary triggers like walnuts or chemo. Perhaps the reason why interferon was so effective in this small study is that there is a viral component involved with other causes. The results of this study could not have been more robust.

The nine patients given human interferon had complete remission of their ulcers within two weeks.

Dr. Hutchison's pioneering spirit greatly aided our understanding of how best to use low-dose oral human interferon in humans. His willingness to experiment with something new seems to be uncommon among physicians, who are generally reluctant to step out of the accepted paradigms of medical practice. Dr. Hutchinson's friendship, courage, and curiosity made it possible to learn from his clinical observations about the use of human interferon in many different diseases. His clinical observations were invaluable as we planned clinical trials seeking FDA approval. It's safe to say that we would probably not have conducted human clinical trials if Dr. Hutchinson had not "pointed the way." Although his observations were not "placebo-controlled," his experiences guided us on dose and in selecting clinical indications. When Dr. Hutchinson died from a stroke in 2004, we lost a valuable friend and colleague.

Because of Dr. Hutchinson's success treating aphthous stomatitis, a study was conducted at the University of Texas in San Antonio. The study tested placebo or 15, 50, or 150 IU of human interferon in fifty-six patients with three bouts of aphthous stomatitis. Treatment was given orally for five days starting at the first symptom of aphthous stomatitis. There was so much variability in the size and duration of lesions, we failed to reach statistical significance, and the paper was not published. I also question whether the five days of treatment were enough to fully benefit from the human interferon, as our treatment period in our previous study had been two weeks.

However, even despite these failings, the data clearly showed the 50 IU human interferon dosage consistently outperformed placebo and other doses.

**

My earliest correspondence with Immuno Modulator Labs (IML) was my letter dated December 10, 1982, to Dr. Jerzy Georgiades, founder of IML, in which I informed him that IML's human interferon appeared to cause weight gain and improve appetite in a sick calf. I began to work with IML and helped them gain approval from the Texas Department of Health for a human interferon product called Agriferon to treat cattle with shipping fever. Later, IML was granted approval to treat horses with human interferon, which they called Equiferon.

The sales of Agriferon and Equiferon from 1985 to 1987 were $1,891,121. That was a lot of money in that time and is a testament to IML's marketing skills. Unfortunately, IML later declared bankruptcy, a not-uncommon occurrence among biotech companies. Many investors do not fully appreciate how long good science can take, or the amount of time required for government agencies to act on promising developments.

Even though the company ceased operations, Dr. Georgiades maintained a keen interest in the oral human interferon technology. That interest eventually led him to Poland, where he began five years of clinical trials on the use of low-dose oral human interferon in hepatitis B, hepatitis C, AIDS, and childhood cancer. Dr Georgiades, like Dr. Val Hutchinson, was a courageous pioneer willing to experiment with a new technology.

The outstanding clinical success of Dr. Georgiades in treating hepatitis B was reported in a 1993 special issue of *Archivum Immunologiae & Therapiae Experimentalis*:

> However, even without clinical manifestations, patients with the chronic form of hepatitis, HBsAg, HBeAg and HBV DNA in sera after human interferon treatment often eliminated the virus, regardless of age. Shortly after virus elimination patients also developed specific anti-HBe anti-bodies. In addition, normalization of various biochemical and clinical variables took place. The above findings let us believe that these patients have been cured and have developed immunity . . . Further evidence is provided by the fact that none of the patients who seroconverted during oral human interferon therapy support this notion.[83]

These results were quite dramatic, but it would be a mistake to claim that interferon cured all the patients with hepatitis B. There were some who did not respond, and as any good clinician knows, the patient and scientific curiosity compel you to keep investigating until you determine the answer.

In 1996, Georgiades and other investigators sought to determine the common patterns of those who responded to interferon and those who did not:

> During therapy of chronic viral hepatitis B (CVHB), some patients treated with natural human interferon alpha lozenges failed to respond. These observations triggered studies aimed to determine whether there are markers predicting patients' response to therapy with nHuman interferon alpha lozenges. In these studies, 32 patients with CVHB were involved: 20 males and 12 females, 16-61 years of age with proven persistent hepatitis B viremia (HBV). Patients were evaluated for clinical, biochemical liver function, and virological markers of disease. During 300 days of treatment of the patients received 75-150 IU human interferon daily in form of lozenges.
>
> The responders to oral interferon therapy were those who had initially alanine amino transferase (ALAT) level higher than 100 U (85.7% cure rate) and weak responses were observed among patients who had an initial ALAT level below 100 U (9.0% response rate). Therefore, ALAT test in patients with CVHB may serve as a predicting indicator of the outcome of human interferon lozenges therapy.[84]

What I love about science is that what others regard as obstacles or failures, Dr. Georgiades regarded as a challenge to more deeply understand the biological process.

Dr. Georgiades was getting some exceptionally positive responses with some of his hepatitis B patients, but not with others. By looking at the underlying clinical markers, he could establish that the key difference between the two groups was their level of alanine amino transferase. Dr. Georgiades and his colleagues in Poland would eventually go on to publish fifteen papers on how low-dose oral human interferon helps control chronic hepatitis.

One of the most important achievements of Dr. Georgiades in Poland was the findings that chronic, active hepatitis B patients responded well to low doses of oral human interferon. The patient clinical responses, using a low dose of oral human interferon, were comparable to those achieved with high-dose injected human interferon. However, he told me that the low-dose oral human interferon therapy was much less expensive, free of adverse reactions, did not require a needle and syringe for administration, and was 1/10,000 the dose of injected human interferon.

No one has worked harder than Dr. Georgiades to make a success of low-dose oral human interferon, and no one is more frustrated by the slow pace of development and acceptance of this technology. Georgiades holds the belief that low-dose oral human interferon provided significant clinical benefits to patients without the severe adverse reactions so common when human interferon is injected in high dosages. He cannot understand how others do not yet appreciate that the low-dose oral human interferon is safe and can be delivered at a price most patients can readily afford.

Besides in Poland, other clinical trials of low-dose oral human interferon in hepatitis B and/or C were conducted in Japan, China, Taiwan, Canada, and the Philippines. The most comprehensive hepatitis C study, using low-dose oral HBL human interferon, was conducted in Taiwan, where 169 patients were enrolled, and was published in the *Journal of Interferon & Cytokine Research* in 2014:

> In 158 patients receiving at least 4 weeks of oral interferon, significantly higher platelet count was found at the end of trial in the 500 IU group (P=0.003). In thrombocytopenic patients, a significantly expedited recovery of platelet was found in the 500 IU group (P=0.002). No drug-related severe adverse events were reported. In conclusion, at 500 IU/day, oral interferon exerted a border-line suppression effect of virological relapse in chronic hepatitis C patients

with mild liver fibrosis. Additionally, it significantly expedited platelet count recovery after the end of peginterferon therapy.[85]

As a result, US Patent No. 9,526,694 titled *Treatment of Thrombocytopenia Using Orally Administered Interferon* was issued on these new and novel observations.

***

In October 1985, I appeared on the cover of *Farm Journal*, which billed itself as "The Magazine of American Agriculture." It was the "Beef Extra" edition. On the cover I'm sitting on a bale of hay, wearing jeans, a blue smock over a white striped button-down shirt, I've got short brown hair, large oval glasses, and a big 1980s-style mustache.

That cover photo showed me holding in my lap an enormous Erlenmeyer flask, which was about 20 percent filled with a brackish fluid, and a plastic tube coming out of the top. It was attached to some sort of pistol-like device with a thin, stainless steel barrel that I'm holding in my hand. Poking out in front of the flask is the head of a young calf. From the expression on the calf's face, I imagined he was thinking, "Don't be so impressed with this guy! He used that device to suck the snot out of my nose! And you call that a scientific breakthrough?"

We were using interferon to treat a condition called "shipping fever," and even then, I had trouble containing my enthusiasm. What can I say? I get excited about my work:

> "It's not a silver bullet," Joe Cummins emphasized to the Texas cattle feeders intent on learning about a new product that could help them fight shipping fever. "It won't keep your calves from getting sick. But it may help them recover sooner."
>
> Cummins was at the podium discussing the merits of interferon, the newest weapon in the shipping fever battle. The Texas A&M, Amarillo based microbiologist is on the leading-edge interferon research in the only state that has received approval to use it on cattle.
>
> Even in his best attempt at giving a cautiously reserved opinion, Cummins can't help showing his enthusiasm for interferon. He tells the crowd of the time he was so excited about the prospects of giving it to a pen of new arrivals that he drove full blast all the way to the feedlot, in first gear.[86]

Maybe getting excited is a failing, but I'm convinced it's optimists who change the world. I think a good scientist follows intriguing hints, not knowing which new worlds may be discovered. And the data were giving us clues that needed to be investigated:

> Cummins first work was with beta interferon, produced from bovine kidney cells in his lab. "We saw some improvement in appetite and in temperature compared with calves that never got the bovine interferon," he notes. "But the effect was short-lived. Continually administering new doses to combat the infection wasn't practical."
>
> So Cummins took a look at an alpha interferon "soup," Agriferon-C, produced from human white blood cells by Immuno-Modulators Laboratories (IML), of Stafford, Tex. The results have been widely variable but consistently positive.
>
> "We've seen peak fever as well as duration of fever reduced," he says. "For example, in a study of cows infected with IBR virus, those receiving Agriferon-C had a 0.8-degree lower peak temperature [105.3 degrees versus 106.1 degrees]. The treated calves also had a temperature above 104 degrees for just three days, versus five days for the control calves. So while we don't stop fever, we do lower the peak somewhat and shorten the duration, which makes the infection less severe."[87]

I think my research still stands up well after more than thirty-five years. Yes, I am optimistic, but I report the data accurately. And even then, I didn't make claims that this was the entire story, or that we had nothing else to learn:

> "This product isn't a cure-all, but we think it can supplement an animal's own interferon as well as traditional antibiotic practices," he says. "What we're really trying to do is flip a switch that will turn an animal's immune system on."[88]

For those of you unfamiliar with shipping fever, it can be a serious condition among cattle, often causing a death rate of around 1 percent of all calves that are shipped. In an accompanying article by a rancher who'd been using the product, he claimed he'd cut the death rate among his animals by more than two-thirds (0.86 percent death rate vs. 0.21 percent death rate).[89] Little did I know that my experience with shipping fever in cattle would lead me to the heart of the HIV-AIDS epidemic.

I think the reader can now understand how I view myself as an average veterinarian with a deep love of animals, a dislike of foolish authority, a few good ideas that might help human and animal health, and that I remember myself as that young man who regularly milked cows.

Nothing prepared me for the controversy that was about to engulf my life for decades.

Did I expect this former Ohio farm boy would end up in the very center of the HIV-AIDS epidemic, proposing interferon as a treatment for that scourge?

And that along the way I'd be written about in publications as diverse as the *New York Times, Newsweek,* and the *Chicago Tribune*? Could I have ever imagined I'd be dealing with ministers of African nations, the Nation of Islam, and top health officials of our government, and the World Health Organization (WHO)?

The answer to all of these questions is no.

I could not have imagined it in my wildest dreams.

Maybe I'm like most people who get thrust into the limelight. One day I'm simply doing my job, following up interesting questions, and then an entire new world opens.

***

In the summer of 1989, I flew to Japan to discuss a license to acquire a natural human interferon made by HBL in Okayama. Mr. Katsuaki Hayashibara and five other men met with me for several hours to discuss my research with oral human interferon and the process used for producing it. I sought a license and supply agreement to use their human interferon for oral administration by offering to pay them more than they were receiving for their injectable human interferon. At the end of the day, Mr. Hayashibara proposed that I take some of their human interferon and test it orally in cattle to see if it worked. I accepted his proposal and took the product in my pocket to Texas.

Experimental light-weight feeder calves were waiting in Texas. Within days, twelve calves were inoculated with IBR virus, and some of them were given the low-dose oral human interferon from Japan. The treatment reduced the fever and resulted in better weight gain, compared to control calves. I faxed the results to HBL and asked again for the opportunity to license their human interferon.

With my mother Nora Mae and my younger brother Jeffrey Charles
in the San Diego area (National City) in 1952.

Me in Long Beach in 1946, wearing my dad's World War II combat helmet.

The barn at the farm in Ohio around 1958 where we milked cows.

Even from an early age, I liked to make a lot of noise. Me and my trumpet in National City circa 1950 before I switched to cello.

Don't I look like a young man who will change the world? In National City around 1952.

Signing papers with Ken Hayashibara in 1990 in Japan, to produce interferon.

Guests from Korea visiting Amarillo Bioscience in 1992.

Have I mentioned I generally prefer animals to people? This was taken in Bushland,
Texas in 1993 at the experimental Texas A&M feedlot.

Visiting with Mr. Kim from Korea. Why is it when any visitors come to Texas, they immediately want to get a cowboy hat?

I prefer jeans and boots, but sometimes you need to dress up and look respectable. In Amarillo around 1986.

On the phone, trying to make deals for Amarillo Bioscience in 1996.

Farm Journal.  Beef Extra.

THE MAGAZINE OF AMERICAN AGRICULTURE ® OCTOBER 1985

Is preg checking
a waste of money?

Chop grain
sorghum for silage

Finish calves
on paper first

The Merc wants
cash, not cattle

New weapons in
the war against
shipping fever

Cover story:

Texans give
interferon
a tryout

Bulk Rate
U.S. Postage
PAID
Farm Journal
Inc.

Farm Journal. 230 W. Washington Square, Phila., PA 19105

Tom Selleck may have graced a lot of magazine covers in the 1980s,
but I think I did it better.

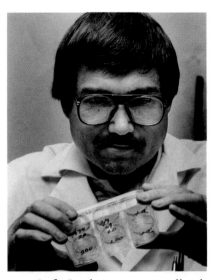 

Left: Looking at some cell cultures treated with interferon at Texas A&M around 1982. Right: Me as a young man milking cows around 1959. Maybe I should have stayed on the farm.

My high school picture as a senior in 1960.

Using long nasal swabs on a cow to get interferon in Bushland, around 1984.

Looking at bovine fetal kidney cell cultures in 1983 at Texas A&M University research facility.

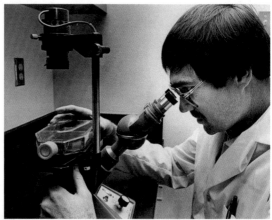

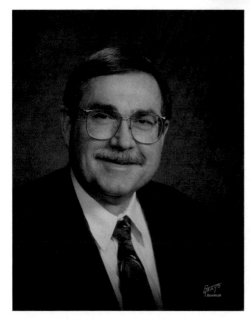

The older scientist, wondering if I contributed anything to the world.

At their request, a license proposal was sent, and I was invited to return to Japan.

When I entered the meeting room in Okayama for the second time, Mr. Katsuaki Hayashibara took me aside and told me that Mr. Ken Hayashibara (HBL's president and his cousin) liked the oral human interferon technology I'd pioneered.

He added that Ken wanted "to invest" and asked, "Would that be possible?"

I was surprised by his statements. I felt I'd just been sharing my work, not pitching them for an investment. At the time I'd been funding my human interferon research with investments from Dennis Moore and other investors in ACC. I felt I barely had two nickels to rub together. I was a science guy, not a corporate money strategist.

I'd been down that road many times before and had failed spectacularly. I couldn't interest any American companies in what I was doing, so I figured that wasn't in the cards.

However, even with my enormous blind spot as to what was really happening, I was able to stammer out, "Yes, that would be possible."

As the discussion went on, I finally asked, "Have you tested oral human interferon?"

"Yes," they responded, and one of the men hurried from the room and returned with their special white powdered sugar preparation (crystalline anhydrous maltose [ACM]) and HBL human interferon that HBL scientists had been testing.

Unlike the sixty companies in the United States I'd approached, who'd rejected my technology without testing, HBL had tested oral human interferon and liked the results. Mr. Ken Hayashibara told me that "corns" came off his feet after he tried oral human interferon.

While in Japan, I agreed to the sale of 10 percent of ABI's stock to HBL at the lowest price I thought my Board of Directors would accept (I chose the option price then held by the colorful T. Boone Pickens). These men had not heard of T. Boone Pickens. But by coincidence that night, he was on Japanese TV because he was in Japan seeking to buy a Japanese auto-parts company. Boone was always such a great media personality that anybody who saw him on television was immediately taken by the tall, plain-speaking Texan. It seemed like fortune was smiling on us.

HBL agreed to grant a license and supply agreement to their human interferon in October 1989, within ninety days of our first meeting. They granted a license and also purchased 10 percent of ABI stock. Surely, this is a speed record for a deal in Japan.

Shortly after we signed agreements with HBL, they completed their Kibi Plant in Okayana, Japan. One must understand the importance of "honor" in the Japanese culture, and although I am not sure our work had anything to do with it, within a few months of signing our deal, their Kibi Plant was honored with a royal visit by the Imperial Highness, Crown Prince Akhito and Crown Princess Michiko, then the Emperor and Empress of Japan.[90]

HBL was best possible partners I could have found. (I mean, before the embezzlement was discovered and they had to shutter their medical research!) HBL worked closely with me and funded (initially, nine million dollars) an extensive research and development plan to test low-dose oral HBL human interferon in human and animal health. They purchased additional stock until their ownership of ABI increased to 44 percent. When ABI went public in 1996, 160 employees of HBL bought our stock in the offering. HBL provided a total of almost eighteen million dollars in support of human interferon research. Their patience, support, and forbearance kept us going for many years. However, it was not to be a permanent situation.

Several years later, it was a profound tragedy that (after 130 years in business) $1.4 billion US dollars were missing from the coffers of HBL, they filed for bankruptcy, and our deal went away.

CHAPTER 7

A Veterinarian Gets HIV-AIDS

In November 1983, Dr. Buddy Brandt, a veterinarian, underwent extensive surgery with resulting complications requiring multiple blood transfusions. By January 1984, Buddy developed pneumonia requiring hospitalization. In retrospect, the pneumonia probably was his first acquired immunodeficiency deficiency syndrome (AIDS)-related illness. Shingles, genital warts, diarrhea, cold sores, mouth ulcers, respiratory infections, and weight loss were chronic problems experienced by Buddy until February, 1986, when AIDS-related complex was finally diagnosed.

In the spring of 1986, Buddy called and asked how he might use oral human interferon to treat a big dog with a tumor. I told him of my experiences using oral human interferon in dogs and advised him as to dose and timing of administration. Because of Buddy's success in treating feline leukemia with Pet Interferon Alpha (then sold in Texas), and because of Buddy's chronic weight loss and depressed CD4+ cell counts (less than 200 cells/cu.mm blood), Buddy experimented on himself with oral Pet Interferon Alpha (low-dose human interferon).

Four benefits from oral human interferon seemed to result as he reported to me:

1. The CD4+ cell count increased from 153 to 319/cu.mm blood.
2. His genital warts regressed.
3. His appetite improved.
4. He gained weight.

The Pet Interferon Alpha treatment was discussed with his physician, who objected to treatment with oral human interferon. This physician told Buddy to discontinue oral human interferon and take the antiretroviral Ribavirin.

During the next eight months of Ribavirin treatment, Buddy experienced weight loss, lost his appetite, and his CD4+ cell count decreased.

At the end of January 1987, Buddy contacted me again and told me he was not treating a big dog but instead was treating himself for AIDS. He then sought my advice on the use of Pet Interferon Alpha for the treatment of AIDS.

He restarted himself on Pet Interferon Alpha oral treatment (at one-tenth the dose taken in May 1986) and claimed he experienced appetite stimulation again.

In February 1987, he gained five pounds.

On February 26, 1987, one month after a previous blood sample, another blood sample was taken and analyzed for CD4+ cell counts, and the count was slightly increased from 189 to 232/cu.mm of blood.

From February to April, his CD4+ cell count decreased from 232 to 210/cu.mm, while his CD8+ cell count increased from 549 to 1050/cu.mm.

From May to June (after taking bovine interferon), his CD4+ cell count rose to 520, and his CD8+ cell count increased to 1352, a strong indicator that his immune system was improving. Another blood sample was tested on June 29 and confirmed blood cell improvement that seemed to correlate with Buddy's clinical improvement.

These numerical increases in his blood counts may not have been meaningful, as sometimes there were natural variations in test results. However, Buddy claims he felt better than he had in three years and attributed his improvement to the oral interferon treatment. He continued working full-time, was active in 4-H, and kept a garden. In other words, he maintained the quality of his life for many months, something he was unable to do before. This information was eventually written up as a case report that Dr. Hutchison and I submitted as a letter and published in one of the most prestigious medical journals in the world, *The Lancet*:

> SIR, - In late 1983, a 40-year-old man underwent open-heart surgery and complications necessitated many blood transfusions. In early 1984 he was admitted to hospital with pneumonia, and this may have been his first AIDS-related illness. Subsequently, herpes zoster, genital warts, diarrhea,

herpes simplex, mouth ulcers, respiratory infections, and weight loss became chronic, and in February 1986, AIDS-related complex (ARC) was diagnosed.

Our patient was a veterinary surgeon. Success with oral human interferon-alpha (IFN-a; pet IFN-a, Amarillo Cell Culture Co) in feline leukemia and his chronic weight loss and depressed T4 lymphocyte counts prompted him to experiment with oral IFN-a. Self-treatment was followed by a rise in T4 lymphocyte count from 153 to 319 /ul, regression of genital warts, improved appetite, and weight gain.

The interferon treatment was discussed with his physician who discouraged its use, and the treatment was discontinued; his T4 lymphocyte count decreased within 6 weeks to 146 /ul. His physician prescribed ribavirin for the next 8 months, and during ribavirin treatment the patient experienced weight loss and anorexia, and his T4 count varied from 146 to 218 /ul. In January 1987, his condition had deteriorated from ARC to AIDS, according to his primary physician.

At the end of January, 1987, he began a course of low-dose (2-4) units/kg body weight per day) oral human IFN-a, and over the past 11 months his appetite has returned, he has gained weight, his herpes simplex and mouth ulcers have disappeared, the genital warts have regressed and recurred, and T4 counts have risen to 210-520 /ul. He feels better than he has for 3 years and has been able to maintain the quality of his life and has not been off work for the 11 months . . .[91]

I hate to say this, but it follows a similar pattern I've seen over the years, especially with respect to low-dose human interferon.

People cannot believe the positive results obtained by an unconventional therapy, so they switch to another medication prescribed by their doctor, only to find their medical condition getting worse. I know nothing is an absolute cure-all, granting immortality, but it certainly seems human interferon should be ranked among the top choices of all possible medical interventions.

Look at what the data showed in this patient.

His T4 lymphocyte count had sunk to 159/ul, then after using interferon it went up to 319/ul. When he went on ribavirin, his T4 lymphocyte count sank to an even lower range of anywhere from 146/ul to 218/ul. The man was dying, and his doctor wanted to keep him on the same protocol, not because it was working, but because it was "approved."

Luckily, he chose to go against his doctor's advice and resumed treatment with oral interferon. Besides the resolution of clinical signs, his T4 counts rose to a range of 210-520/ul, and he felt better than he had in years. In the concluding section of our letter, we wrote:

> Dolei et al. have suggested that AIDS patients could gain more advantage from repeated low doses of interferon than from large amounts. The patient described here has taken no other treatment for AIDS for the past 11 months. The benefits include low cost (less than $50 per month), the absence of known toxicity from low dosage, and ease of administration.[92]

Because of our success with Buddy, we were eventually able to get a meeting with representatives of T. Boone Pickens, the wealthy and colorful Texas oilman who was interested in funding a larger trial of oral human interferon for HIV-AIDS.

Buddy eventually died in November 1990 when he failed to recover from complications related to more heart surgery. Buddy had become a good friend and his death was a great personal and professional loss to me.

Many others also felt the loss of Dr. Buddy Brandt very deeply.

A few years later a young woman, Arianna Nesbit, an American exchange student living in New Zealand, asked that Dr. Buddy's name be put on the country's AIDS quilt. She said he'd "been like a father to me," and in her tribute to him she wrote, "Dr. Buddy was born on 9 April 1943 and died on 16 November 1990. He graduated from Texas A&M University with a degree in veterinary medicine and loved animals and children. He loved horses, dogs, birds, mules, and peacocks. He was a man that had a very big heart and cared about everyone."[93]

In his section of the quilt is a hilly landscape with a tree and a bright yellow sun. In the left corner are the initials of Texas A&M University, along with the AGGIES name, the mascot of the school. The top leaf of the tree has the words "With love from," and on the ten other leaves of the tree are the names "Arianna, E.M., Matt, Eddie, T.J., Tina, Kelly, GGG, Eva, and Sean." Written on the sun is the name "Dr. Buddy, April 9, 1943–November 16, 1990. He is still living in the hearts of those who love him." He touched many lives.

We do not forget our heroes.

***

In 1988, while Dr. Buddy was still alive and practicing veterinary medicine, I was visited by Dr. Alan Young and Ms. Casey Burns. Dr. Young was an entomologist in Kenya working then primarily on East Coast Fever (ECF), a highly fatal tick-transmitted protozoan disease of cattle. Dr. Young had heard about Agriferon and wanted to test low-dose oral human interferon in ECF. We discussed research, and I gave him Agriferon to take home to Kenya. Much to his surprise, oral human interferon was lifesaving in calves with ECF.[94]

In March 1989, I flew to Kenya to visit Dr. Young and to see his research facilities. I was delighted to learn that excellent, high-quality animal research was being conducted in Nairobi at the International Laboratory for Research in Animal Disease (ILRAD) and at the Kenya Agricultural Research Institute (KARI). While in Nairobi, I was introduced to Dr. Davy Koech, the director of the Kenya Medical Research Institute (KEMRI), a modern facility built by the Japanese as a gift to the Kenyan people. As the head of KEMRI, he was interested in any potential treatment for AIDS. I shared with him our observations regarding Buddy Brandt and our work with feline leukemia and oral human interferon.

In July 1989, I returned to Nairobi and shared the preliminary data from our AIDS study in Amarillo. At that time, I asked Dr. Koech to test oral human interferon in AIDS patients in Kenya. He readily consented. Based on our preliminary data on AIDS patients, Buddy Brandt's story, and the successful treatment of cats with feline leukemia, Dr. Koech and his medical director, Dr. Arthur Obel, boldly pushed KEMRI's institutional review board for approval to treat Kenyan AIDS patients with natural oral human interferon from Japan.

The natural human interferon tested was a mixture of IFNα 2b, IFNα 7, and IFNα 8 from HBL in Okayama, Japan. This Ohio farm boy had ended up as an associate professor at Texas A&M University, flown to Kenya, and convinced a Japanese company to partner with me.

When I think of all the headaches this eventually caused me, I often wonder if I should have just stuck to milking cows or spaying dogs and cats.

***

In October 1989, ACM powder containing HBL human interferon was brought to Kenya so Dr. Obel and Dr. Koech could start treating AIDS patients. After returning to Amarillo, I telephoned Dr. Obel and asked how his patients were doing.

He told me that the first patients were experiencing nausea.

Because of the nausea, I advised Dr. Obel to "cut the dose," and he reduced the treatment to once a day and reduced the amount of a single dose from "250 IU" to "200 IU." I never knew what dose was given in Kenya because the powdered ACM/human interferon was given in small volumes measured by Dr. Koech with the clip off a Bic pen. Later, we learned that the ACM powder would keep the HBL interferon stable if the moisture content stayed below 2 percent. Exposed to the air, the ACM powder attracted moisture. Therefore, we never knew how much human interferon activity was in the doses of powder given in Kenya.

My best guess is that the activity was considerably less than originally thought.

After Dr. Obel reduced the dose, he reported clinical improvement in his patients. Patients became asymptomatic, and his patients reported improved appetite and sexual drive. The laboratory reports of improved blood counts were impressive. The "CD4+ lymphocyte counts," then the most accepted measure of progression of AIDS, were reported to improve from very low counts to normal.

I flew back to Kenya to visit Drs. Obel and Koech to review the situation, interview doctors and patients, and monitor the patients' records and laboratory reports. I was disappointed to discover the records were incomplete. The laboratory reports and clinical data (weight change, symptom reports, etc.) were missing, and it was generally hard to ascertain from the records exactly which benefits were occurring.

But the enthusiasm of Dr. Obel, who treated patients, and what I saw with my own eyes of the patients were encouraging.

Over the next few weeks, Dr. Koech completed a data set on forty-two patients. The progress in Nairobi was reviewed from Amarillo as the biweekly laboratory reports arrived by fax. So impressive were these reports that I returned to visit Nairobi.

Again, the record keeping left much to be desired. I tried to remember that I was a guest in Kenya, and they had their own way of handling records, frustrating as it was to an "outsider."

Eventually I was able to get the records into good shape, and we published an article in the journal *Molecular Biotherapy* in June 1990. In the opening we reviewed human interferon, as well as its medical acceptance:

> Human interferon alpha is a family of leukocyte-derived proteins with immunomodulatory, antiproliferative, and antiviral properties. Recombinant and natural human interferon previously were approved [by the FDA] for

the treatment of hairy cell leukemia, condyloma acuminate, and Kaposi's sarcoma.[95]

This is another testament to how oral human interferon can modulate the functioning of the immune system in a positive direction, reduce the growth of cancer, and protect against viruses. The safety of natural human interferon, as well as that produced by recombinant procedures, has been approved by the US FDA for multiple conditions.

I think the approval for Kaposi's sarcoma is particularly relevant, as that is one of the first conditions to appear among early sufferers of AIDS patients. Before the HIV-AIDS epidemic, Kaposi's sarcoma was most commonly found in elderly Italian men from the southern tip of Italy. The condition causes purplish cancerous skin masses to appear on the face, lymph nodes, and other areas of the body. Why were young, healthy gay men in the United States coming down with this condition?

It was because the HIV virus was disrupting the proper functioning of their immune system:

> The critical feature of the loss of immunocompetence in HIV-1 infection is the depletion of the helper-induced subset of T-lymphocytes which express the CD4 molecule. This depletion of CD4+ cells is associated with impairment of function in multiple components of the immune response and results in profound immune dysfunction.
>
> The HIV-1 selectively infects cells expressing the CD4 molecule, resulting in quantitative and qualitative defects in CD4+ T-Lymphocyte function in patients with AIDS. It has been demonstrated that the CD4+ T-lymphocyte function in patients with AIDS. It has been demonstrated that the CD4+ T cell is the predominant cell harboring HIV-1 in the peripheral blood of infected individuals. These cells actively express HIV-1 as determined *in situ* hybridization to detect viral RNA, immunofluorescence to detect viral antigens, and limiting dilution of co-cultures.[96]

I think it's probably helpful at this point to explain a few things about the above passage. Somebody considering Koch's postulates, first promulgated in the 1890s, would read that section and say that HIV isn't a virus that causes disease.

According to Koch's postulates, the virus would need to be isolated from a sick individual and when injected into a healthy organism would always cause the disease. That's been proven to be wrong due to our evolving understanding of genetics. There can be "elite controllers," those with a different genetic profile, who are immune to the virus. Arsenic is a poison, for example, but the genetics of different individuals require higher dosages if you want to kill them.

I advise you to be wary of any scientific principles developed before Orville and Wilbur Wright first took flight at Kitty Hawk, North Carolina, in 1903. It's not so much that they're wrong, but they're incomplete. Every hundred years or so, there are usually some adjustments to any generally accepted scientific principles.

Just look at how our understanding of gravity has changed since that apple supposedly fell on Isaac Newton's head. We know that HIV causes its damage by altering the functioning of the immune system.

The next part of the discussion section described how the success of any HIV-AIDS intervention was measured, and what we found in our 1990 published study:

> Survival has been used as an indicator of success in the treatment of AIDS patients, but other criteria are used to assess clinical response. Measures of macrophages, ADCC, NK cells, Beta-2-microglobulin, Neopterin, and DTH have been proposed to help assess efficacy of various treatments. Clinical measurements of weight, temperature, fatigue, and diarrhea have been used to determine progress or regression in AIDS patients. Systems to assess staging, or to calculate a clinical index of disease, have used many of these measurements. However, CD4+ cell counts during treatment remain the most important predictor of survival.
>
> The CD4+ cell count increased in all but 4 of the treated patients; the average increase was 291.1 cells/cu.mm3 after 2 weeks of natural human interferon treatment. Concurrent with increases in the CD4+ cell count were increases in the KPS [Karnofksy Performance Status – A measure of a person's ability to perform the daily activities of life. The range is from 1-100, with 100 being normal.] Patients experienced dramatic symptomatic relief and all of the patients except 3 had a KPS of 100 after 2, 4, or 6 weeks of human interferon treatment. The average weight gain of patients after six weeks of oral human interferon therapy was 4.6 kg [a little more than 10 pounds] and is in contrast to loss of appetite and other toxic effects associated with high dosage parenteral usage of human interferon. The low dosage and oral administration of human interferon differ in comparison to other human interferon treatment regimens that have been tested in AIDS and HIV-1 seropositive patients.[97]

To put it in plain terms, the best indicator of survival for an HIV-AIDS patient is their CD4+ cell counts. The average increase in their CD4+ cell count was 291.1. All but three of the forty patients were able to get a perfect score of a hundred on the KPS, which measured their ability to perform the routine functions of daily life.

And in six weeks, the average weight gain per person was a little more than ten pounds, a remarkable result considering the disease causes a loss of appetite and a slow wasting away.

The final revelation of the paper is perhaps the most shocking. While we suggested that caution was warranted in assessing the positive changes in the CD4+ cell populations, and whether changes would be long-lasting, we believed it was worthy of further study. But the biggest bombshell of the paper was what reportedly happened to eight of the patients:

> Four patients turned negative for both ELISA and Western Blot by the second week after treatment, and another 4 patients turned negative by the fourth week. These individuals have since remained negative during subsequent follow-ups and have continued to be asymptomatic. This is the first observation where there has been "seroconversion" on both ELISA and Western Blot and clinical improvement resulting, apparently, from therapy.[98]

Had we achieved at least a functional cure for 20 percent of our study participants? It was difficult to come to any other conclusion, although we would certainly need to engage in further studies. Perhaps the virus was hiding out in hidden reservoirs of the body, but if it stayed hidden, would it ever do any damage?

I don't think it's overstating our findings to say they were among the most exciting in the history of the HIV-AIDS epidemic. Our AIDS data would be rejected, just as Soloviev and his flu data was rejected in 1969.

Now, I understand if some readers might be doubtful of my claims. So, we did some experiments in Africa and got these wild results. One might think results that impressive would surely have been reported in the pages of the *New York Times*, maybe with an accompanying boost from a top AIDS scientist like Anthony Fauci, head of the National Institutes of Allergy and Infectious Diseases (NIAID).

In fact, Dr. Fauci and his colleagues reported similar results a year earlier, as reported in a *New York Times* article from August 15, 1989:

As AIDS patients and their advocates go to the ends of the earth to obtain unapproved drugs, an approved drug that is already on the market has languished in neglect, even though it shows some signs of effectiveness in delaying the onset or progress of the disease.

The drug, alpha interferon, has been tested in several small studies in recent years. It has been deemed promising by Dr. Anthony S. Fauci, director of the NIAID, who leads the Federal AIDS research effort, and by prominent clinical investigators like Dr. Jeffrey Laurence of Cornell University Medical College in Manhattan, Dr. Donald Abrams of San Francisco General Hospital and Dr. Jerome Groopman of New England Deaconess Hospital in Boston.

The drug should not be oversold, experts warn. It has not been tested in large-scale studies and has not been proved to extend the lives of people with AIDS. But data from preliminary studies indicate that the drug may prevent the onset of disease in some people infected with the AIDS virus and that it may slow the disease's progress in those who already show symptoms.[99]

Now this part of the article just simply sums up the outline of what it's going to show. It also had the parade of experts telling you this medication is something that should be considered, including Dr. Fauci of the NIAID and prominent researchers from Cornell University, San Francisco, and Boston. And of course, it has the expected warnings that we need more research. I agree with all those precautions. Now, let's examine the results:

In 1990, AZT was the only drug approved for the treatment of acquired immune deficiency syndrome, but as many as half of all patients cannot tolerate the recommended doses because the drug damages their bone marrow.

In the study, 22 men infected with the AIDS virus took low doses of the two drugs for 12 weeks and the drugs apparently worked together to inhibit the viral infection. For example, six men had viral proteins in their blood when the study began, but those proteins disappeared in three of them. Six of the 22 men no longer had viruses that could be isolated from their blood.[100]

Let's look at the problems. AZT, while effective, showed the dose levels used could damage the bone marrow in about half of the patients.

Big problem if you want to continue living.

But if you combined the AZT at a low dose with interferon-alpha at a low dose, you got a significantly different result. Viral proteins disappeared in half of those who were most sick, and the virus itself disappeared completely from the blood of twenty-seven percent of the patients:

> Dr. Fauci said the growing evidence that interferon is effective against AIDS patients who could not take AZT because of its side effects should consider taking alpha interferon instead of or in combination with AZT.
>
> "I believe that alpha interferon is very promising," he said. "Independent studies in the United States and Canada have clearly shown it is effective in Kaposi's sarcoma and that it had a significant effect" against the AIDS virus.
>
> "I am a little bit puzzled about why it hasn't caught on," he said, especially when AIDS patients and their advocates were "getting excited about drugs that barely had effects in the laboratory and had no evidence of clinical effects."[101]

There you have the doctor leading the government's response to HIV-AIDS proclaiming, long ago, that this therapy deserved serious consideration. And as I keep saying, dosage matters. The body naturally has low levels of interferon during an infection:

> The puzzle of why alpha interferon has met with such an indifferent response illustrates the roller-coaster politics of AIDS drugs, some experts said. Alpha interferon has never become, a "drug du jour," said Dr. Groopman of New England Deaconess. And so it has languished while other less promising drugs have flourished in the growing underground market for AIDS drugs.
>
> Advocates for AIDS patients said they had not pushed for the drug in the way they had requested other drugs. "I do think that early on, alpha interferon got a reputation in the community for having side effects," said James Eigo of ACT UP, the AIDS Coalition to Unleash Power. "It may be a while before it woos the community back."
>
> The most recent studies use much lower doses, which have sharply reduced the problem of side effects.[102]

Time after time I stand amazed at people who think they can improve on what nature already provides. We should try to follow what nature shows us to be healthy, rather than try to supercharge health in a way that was never intended. The *New York Times* article continued:

> Interferon's image problems date to the 1970's when some cancer researchers, including Dr. Mathilde Krim, a founder of the American Foundation for AIDS Research, enthusiastically promoted the drug as a potential cure for cancer. But it proved immensely disappointing. It was ineffective against the

more common cancers and was found to be effective mainly against hairy
cell leukemia.

"The disappointment of the anticancer treatment has carried over,"
Dr. Krim said. "It created a negative feeling about alpha interferon among
physicians."

But in 1983, AIDS researchers started testing the drug in patients with
Kaposi's sarcoma, Dr. Krim said. They started with patients with large tumors
and gave very high doses of the drug, which can cause malaise and fevers and
flu-like side effects.

About half the patients responded. But, Dr. Krim said, "No attempt
was made to work with very early disease at lower doses." As a result, many
patients decided that the side effects of the drug made it unacceptable.[103]

As a scientist, I have to say those last paragraphs are disturbing on so
many levels. Was it a blind spot in believing that if a little is good, more
must be better, or part of a well-orchestrated plan to ensure human interferon never got a fair trial? It does seem like that has been a trend in the use
of human interferon in past trials in humans. And it certainly contradicts
the results found in many other studies where low-dose human interferon
was positively used for a number of conditions.

I wish I could peer into the souls and minds of the researchers to answer
this question, but I can't. I can only criticize their shortchanging of the
possibilities.

A scientist is supposed to be rational, testing various alternatives so they
understand all the data. There should have been three more parts to that
investigation to determine if interferon might provide a clinical benefit.

The first was to take a group with large Kaposi's sarcoma tumors and
give them a low dose of human interferon.

The second would have been to take a group with newly formed, relatively small Kaposi's sarcoma tumors and give them a high dose of human
interferon.

The third would have been to take a group with newly formed, relatively
small, Kaposi's sarcoma tumors and give them a low-dose human interferon.

When that investigation was finished, there would have been four total
groups. Those with Kaposi's sarcoma and large tumors, and those with
Kaposi's sarcoma and small tumors, and each one of those two groups
would have been divided into those who received high-dose or low-dose
human interferon.

That would've established four distinct groups that could be evaluated for potential efficacy and dosage.

A middle school science teacher would've known enough to set up that kind of an experiment to test a new compound. Why didn't our government's top scientists?

> But Dr. Krim and others, including Dr. Laurence, said much lower doses of alpha interferon should be tried much earlier in infection, and as a treatment for the AIDS virus itself rather than just for Kaposi's sarcoma. Lower doses cause no noticeable side effects, Dr. Krim said, and the cost of the drug would only be about $30 a week.
>
> Dr. Laurence added that the AIDS virus is very unlikely to become resistant to alpha interferon because of the complex way the drug works. But the virus can become resistant to AZT and, presumably, other drugs that work in similar ways.[104]

Maybe now that you have an example from the *New York Times* of the criticisms I've been making in this book, you're starting to believe this old veterinarian has some valid points.

How much suffering was caused by HIV-AIDS due to this failure to conduct an even reasonably competent investigation of human interferon? My mind staggers at the possibility. The HIV-AIDS activists were among the loudest and most vocal critics of the 1980s and 1990s as the epidemic rampaged through society.

In retrospect, I'm not sure they were angry enough.

***

Thomas Jefferson is alleged to have once said, "When injustice becomes law, resistance becomes duty." Whether he spoke those actual words, or something like it, is open to debate, but there's no denying the bedrock American principle behind it. In that Jeffersonian tradition of valuing human need above any law or government, I give you an article from 1990 about Ron Woodroof.

He's remembered as the real-life main character in the film *Dallas Buyer's Club*. Woodroof contracted AIDS and, when first diagnosed, was given mere weeks to live. He aggressively sought out the newest drugs and medicine, even if they were not yet legal, and ended up living seven more years and helping thousands of people. I never had the pleasure of meeting Ron Woodroof, but he knew my work, and I greatly admired him:

> The thin man in gray approached customs with contraband – 36 vials of a life-renewing drug packed in dry ice in a black leather briefcase – bound for Dallas.
>
> It should have been easy to slip out of Tokyo, but there were problems. A conspicuous frosty patch had condensed on the surface of the briefcase, and smoke seeped through its sides as dry ice evaporated.
>
> Ron Woodroof, an experienced smuggler, moved fast.
>
> He slipped the vials into his pockets and popped open the briefcase for Japanese officials, understandably suspicious of smoking luggage.
>
> "Why are you carrying dry ice around the world?" one asked.
>
> "Would you believe," Woodroof replied, "that it's a fetish of mine?"
>
> Victorious, he boarded his plane. Within two weeks, in April, he had begun shipping oral alpha interferon to hundreds of U.S. men and women dying of AIDS.[105]

Although I never met Ron Woodroof, there were times when I felt almost like an outlaw because of what I was doing. Robin Hood took from the rich to give to the poor. Were Woodroof and I committing the crime of giving hope to people and stealing market share from the pharmaceutical companies?

I always think it's necessary to get third-party, unbiased opinions if you want to establish the truth of anything. Ron Woodroof became famous for his fearless stance to get anything that might work for HIV-AIDS. And remember, he had no reason to endorse my work:

> Woodroof, a 40-year-old former construction worker who was diagnosed with AIDS in 1987, takes the interferon daily, as does his staff.
>
> "I feel like a million bucks," he said. "When you take the alpha interferon, it makes you feel like you did 10 years ago – you have energy, you regain your appetite.
>
> Whether this will continue, I don't know. But I'm damn well impressed."
>
> The results obtained by members of a U.S. buyers clubs so far have reflected findings of an African study using the drug and method of treatment for AIDS developed by Dr. Joseph Cummins, an Amarillo veterinary researcher.[106]

I also have to say that Woodroof was reported to be a very colorful character, even supposedly bringing a gun into his doctor's office, although that's not so uncommon in Texas as you might think. The article about Woodroof then went into detail about my work:

> It was Cummins who thought of administering alpha-interferon orally in small doses, rejecting the usual practice of injecting massive amounts – which often yields a host of painful side effects.
>
> Cummins successfully treated viral diseases of cattle and cats before theorizing that oral alpha interferon, an anti-viral protein, may be used to relieve AIDS symptoms.
>
> With financial aid from a Japanese pharmaceutical company, Cummins tested his treatment on AIDS patients in Kenya.
>
> The results were so encouraging, even Cummins was reluctant to accept them. About 40 men and women with AIDS gained weight and lost symptoms such as fever, fatigue and diarrhea.[107]

I was appreciative of the article for giving me the appropriate credit for the idea of administering small doses of alpha interferon, as well as my successful treatment of cattle and cats suffering from viral diseases.

As much as I'm enthusiastic about alpha-interferon, when those first results came in, even I was skeptical.

Near the end of the article, it went on to describe how alpha interferon was being successfully used by members of the Dallas Buyer's Club and what they were discovering in the course of treatment:

> Today, says Woodroof, about 400 members of the Dallas Buyer's Club are using oral alpha interferon. It costs each patient $1 a day – pocket money for AIDS sufferers accustomed to exorbitantly expensive, U.S. manufactured drugs.
>
> The drug is taken with an oral syringe, swished about in the mouth for three minutes, then swallowed. It tastes salty.
>
> Woodroof has received testimonials from around the country praising the treatment, though some problems have surfaced.
>
> One is dosage levels, which Woodroof originally duplicated from Cummins Kenya study, calling for two international units of alpha interferon per kilogram of body weight.
>
> We immediately start by cutting back somewhat," Woodroof said.[108]

I know Woodroof didn't have any sort of medical training, but science shouldn't be so difficult that a regular person can't understand how it can be practiced.

You try something to see if it works, and document what the side effects might be. The trick here is to use less, not more, interferon.

At that time, everybody knew the inevitable consequence of AIDS was a slow and painful death. Time to get creative, right? What have you got to lose?

I also must express my respect for Woodroof giving the alpha-interferon to AIDS patients, watching their reaction, and then adjusting. That is what any good scientist or doctor does. You follow the patient. That's why it's called the "practice" of medicine, rather than the infallible rule book of medicine. Each patient and each condition must be evaluated according to its own unique characteristics. And different groups might get different results for reasons that might be unclear at the time:

> The PWA Health Group in New York has had less success, according to Garance Franke-Ruta, a spokeswoman.
>
> "We've heard an incredibly varied array of reports," she said. "We've had people with improved T-cell counts (an indication that the body is rebuilding its immune system), as well as some with no change.
>
> Recently, we've heard more good reports than bad."[109]

That's what's supposed to characterize the practice of medicine. All of us are in the dark about what will work best in every situation. No medical intervention comes with an absolute guarantee.

It comes with a probability, a road map of potential problems, and how to take actions to lower the risk as much as possible.

But did you notice how even the PWA Health Group noted promising results in some people, including the "improved T-cell counts," which are an indication that the body is "rebuilding its immune system"?

I can't conceive of a more beautiful sentence in all of medicine than "an indication that the body is rebuilding its immune system."

It speaks of a return to an original state of health after a deadly pathogen had rampaged through the body.

And I can't help but think of Ron Woodroof's observation that the dose I was suggesting seemed too high, and he immediately started with a lower dosage. I wonder if our African populations, although infected, might have had a much more robust set of T-cell counts, suggesting why they tolerated the higher dosage.

I don't have their numbers so I can't be certain, but logically it makes sense. Woodroof didn't believe in stopping at any point if you were saving human life. I don't, either.

As long as there is life, there is hope.

CHAPTER 8

Interferon Disaster

There's a concept known as "Hanlon's Razor," which in one version says something along the lines of "Never ascribe to conspiracy that which can be equally well explained by incompetence, stupidity, or foolishness."

I can state as a fact that if human interferon had been approved, it would have been an inexpensive and safer AIDS medication than many of those that eventually made their way onto the market and made billions of dollars for the pharmaceutical industry. But do I think the pharmaceutical companies alone killed, or perhaps delayed, the promise of human interferon?

No, I do not.

I believe a significant contributing factor to the poor reputation of low-dose oral human interferon today is that many individuals confused their own personal ambitions and biases with the need to save human life. That's it. Politics got in the way of science. People are always surprised when you make a claim about some recent incident in which some form of politics affected science. But when one goes into the past, the examples are everywhere you look.

The promising low-dose human interferon results reported by Soloviev in 1967 and 1969 (see Chapter 3) were rejected by "interferon experts." The low dose use of interferon is still rejected fifty years later.

We mistakenly believe we have evolved morally, spiritually, and intellectually from our ancestors of just a few short centuries ago. The sad fact is we have not. If the American Medical Association today said that bloodsucking leeches were a good treatment for disease, probably most of the medical profession would willingly comply without doing any further research.

Okay, now it's back to Africa, and how things went off the rails.

In conversations with Dr. Koech, I congratulated him on his impressive laboratory and clinical data and urged him to conduct a blinded, placebo-controlled clinical trial. After all, the patients in his study knew they were receiving human interferon.

Perhaps, I argued, these patients were benefitting from a placebo effect. Dr. Koech agreed to conduct a blinded, placebo-controlled study.

HBL prepared placebo and different doses of their human interferon for such a trial. By January 1990, HBL prepared thousands of individually foil-wrapped lozenges containing placebo (white foil), 2 IU per lozenge (blue foil), 20 IU per lozenge (yellow foil), or 200 IU of human interferon per lozenge (green foil), and they were delivered to Kenya. This is not the way blinded clinical supplies should be presented.

All dose forms should look the same so patients and doctors could be truly "blinded."

When I left Nairobi in early January 1990, I hoped Drs. Koech and Obel were going to give the recently delivered free lozenges to AIDS patients so we could all learn if their earlier observations were reproducible. Looking back, I should have offered them financial support for the study, not just free lozenges. Unfortunately, the eventual disposition of these lozenges became a mystery.

Apparently, some of these placebo and active lozenges became commercially available. Even now, thirty years later, it's painful to revisit these memories, as I believe we were close to something that would have been an enormous benefit to mankind. My local paper, the *Amarillo News Globe*, accurately summed up the state of my work this way in April 1990:

> Cummins, who served at the Texas A&M University Research Center in Amarillo for six years as an associate professor of veterinary microbiology, placed seven patents on his oral interferon breakthrough. The federal government issued four of his patents, three are still pending.
>
> The Japanese and Dallas oilman T. Boone Pickens both invested in Cummins' medical research and development company. The Japanese company that manufactures the interferon pills owns 36 percent and Pickens' Mesa Limited Partnership owns 11 percent.
>
> "Cummins is close enough to proving his theory that people are paying attention," said Stephen Gens, who observes medical and scientific research in Amarillo as President of the Amarillo Medical Center. "He is onto something."[110]

In November 1989, Dr. Koech made a public announcement that a new AIDS treatment had been discovered; he named the treatment KEO89. At the tenth anniversary of the founding of KEMRI, in February 1990, Dr. Koech announced that the new AIDS "wonder drug" was to be called KEMRON®.

The Kenyan press started carrying inaccurate claims that KEMRON® was "manufactured in a secret Nairobi laboratory" and that KEMRON® was "manufactured" and "invented" by Dr. Koech and the country of Kenya was going to benefit from royalties and recognition of this Kenyan invention.

I telephoned Dr. Koech to complain about the inaccurate press reports. Dr. Koech told me not to worry, that it was the nature of the press to get carried away with a story. "You know how the press is," he said. What I did not know at the time was that Koech was married to the daughter of the Kenyan president. Besides national pride being at issue, was there a little family pride, as well?

In the spring of 1990, KEMRON® was selling for forty dollars (US) per dose in Africa. Some of the lozenges delivered free for the double-blind, placebo-controlled study were being sold. The stories of KEMRON® sales and the claims in the press of an "African cure" for AIDS cast a cloud of suspicion over everything we did in Africa.

Dr. Koech later explained that he did not want to give placebo to any patient because it was "unethical." We never received an accounting of what happened to the foil-wrapped lozenges given to KEMRI.

In the spring of 1990, Dr. Koech told me that WHO wanted to test KEMRON®. After I told him the treatment code, he delivered the green foil-wrapped lozenges to WHO for a "multicenter" clinical trial of four weeks' duration.

Without my knowledge or input, a hasty WHO trial was organized.

One hundred and eight patients were treated in five different countries in Africa (Ivory Coast, Zimbabwe, Kenya, Cameroon, and Congo). The trial was designed to assess the patients after two and four weeks on KEMRON® treatments. The green foil-wrapped lozenges were later assayed at 183 IU per lozenge; the WHO patients all received that once daily dose for four consecutive weeks.

WHO reported to the press in May 1990 that the results were "inconclusive" and "did not replicate" the beneficial reports from Kenya. However, the written report by WHO actually stated that some of the initial clinical signs and symptoms (oral candidiasis, fatigue/weakness, appetite loss,

insomnia, night sweats, dysphagia, cough, diarrhea, pruritus) were amelio-
rated by the end of the study in over 60 percent of the patients.

I now believe that a six-month treatment, not a four-week treatment, is
required to produce a benefit in an AIDS patient.

You can imagine my surprise when, in February 1991, Dr. Koech
sought patent protection for his own human interferon therapy. This is
how it was reported in the *City Sun*, an African-American paper published
in Brooklyn:

> Some months after Kemron was being touted as African-invented, Dr. Joseph
> Oliech, Kenya's director of medical services, admitted in Parliament that his
> claim the KEMRI's director, Dr. Davy Koech, was the sole inventor of the
> drug was not true.
>
> But Koech, through scientific research, did develop a new methodology
> for the use of low-dose oral alpha interferon, the patent lawyer said, and his
> formulation of Kemron differs markedly in its chemical composition from
> other forms of alpha interferon already known and in use.
>
> "Substantially the same is not chemically identical," said KEMRI's attor-
> ney. "Every little nuance, every little chemical modification, can make a big
> difference. If your particular combination has a particular [chemical combi-
> nation] such that it is superior, that combination is patentable."[111]

Are you getting all this lawyer nonsense? I took time to try and help
Africans suffering with HIV-AIDS, went through the recognized officials
of the Kenyan government, coordinated with my Japanese partners, and
failed. I was deeply disappointed. The writer of the *City Sun* article did the
courtesy of covering my side of this dispute:

> KEMRI's application for a U.S. patent more than likely will meet some resis-
> tance from at least one source – Dr. Joseph Cummins, an immunologist at
> Amarillo Cell Culture Inc. in Texas.
>
> Koech had worked closely with Cummins, who had isolated an animal
> virus that behaved much like one associated with AIDS and had developed
> an antibody for it. Cummins' close associate, Dr. Alan B. Richards, said that,
> about a decade ago, after Cummins had noted significant results in treat-
> ing feline leukemia, with low dose alpha interferon, he began to tailor his
> technology toward the application of the drug in cases of acquired immune
> deficiency . . .

"Joe had already been involved with the treatment of AIDS using low-dose oral interferon before he contacted Koech," Richards said, "but nobody was willing to test it because it was too unorthodox."[112]

I really appreciated Alan's defense of our work and the praise for Dr. Koech to test an unorthodox and unproven therapy. As I've said before, I consider myself an Ohio farm boy who made good in life and just wanted to help animals, and who later realized I might help people, as well:

However, *The Weekly Review*, a nationally distributed magazine in Kenya, reported:

"Koech was the principal investigator in the study. Cummins was a basic immunologist, bringing to the team his experience in veterinary science and his theoretical conceptualizations."[113]

Sometimes I wish I'd just stuck to animals. Dogs usually growl before they bite. Apparently, unlike humans, dogs don't have anything like national pride, hubris, or arrogance.

Then, in America, the Nation of Islam got involved to support the Kenyan government's claims. This is how it was reported in the journal *Nature* in October 1992 in an article titled "Racial Tensions Entangle NIH in Dispute over AIDS Drug":

Under pressure from the Nation of Islam and other African-American activists and political groups, the National Institutes of Health (NIH) is reconsidering a report issued earlier this year rejecting Kemron and other interferon alpha-based drugs as useful treatments for people with AIDS. NIH officials concede that they were politically naïve when they evaluated the drugs—often touted as an "African AIDS cure" because some early clinical trials were conducted in Kenya—in the same way as any other experimental therapy without accounting for the desperation and suspicion of those affected by the AIDS epidemic.[114]

Can you understand how troubled I am by this article? I think as human beings we're hardwired to pick a side in any dispute. But what happens when both sides presented are wrong? It's called the "false choice," and that's what was presented in this article.

And what's up with this lame excuse by the NIH that they were "politically naïve when they evaluated the drugs"? What does that even *mean*?

Nearly thirty years after the article was published, I'm still confused by that sentence. I hoped science was free of politics.

I've never heard of conservative or liberal data, just data. Now, you can argue over the interpretation of data until the cows come home. But the data are the data.

Here I was simply trying to help people, and I was getting my reputation trashed because of it. At least *Nature* gave a fairly accurate account of my work:

> Interferon therapy was developed 20 years ago by Joseph Cummins, a Texas veterinary microbiologist and president of the Amarillo Cell Culture Co. Inc. as a treatment for respiratory disease in cattle. Cummins has conducted clinical trials of interferon, with varying success, on a number of human and animal diseases. Earlier this year, he received approval for interferon-alpha clinical trials on AIDS patients in the United States.[115]

If there are detractors who doubt me, or my work, that's fine. I'm optimistic. I'm a veterinarian. But you'd also have to admit I report the data accurately, and in the veterinary field I was so well respected I became an associate professor of veterinary medicine at Texas A&M University. Accomplishing no small feat, I followed standard procedures in order to seek federal approval for the use of human interferon in human patients.

The *Nature* article also gave an accurate account of what I'd experienced in my interactions with the Kenyan medical establishment:

> In 1989, Cummins traveled to Kenya to test interferon alpha in cattle diseases. He met Koech, who was studying AIDS therapies and had seen a 1986 paper by Cummins on the use of low-dose interferon alpha to treat feline leukemia, a disease that is related to AIDS. Cummins showed Koech data from a trial he had started the previous year in Texas of interferon on HIV-positive patients. He then provided Koech with interferon-alpha powder from Japan and Koech began uncontrolled clinical trials on Kenyan AIDS patients.
>
> After a few weeks Koech reported dramatic improvements and announced at a press conference that he had discovered a secret AIDS treatment he called KEO-89. In February 1990, he announced that the drug was oral interferon alpha and would be marketed as KEMRON.[116]

Most of our relationships involve a mix of good and bad situations, and we do our best to navigate them. Dr. Koech had been studying AIDS therapies and was diligent enough to read my 1986 paper so he could discuss it with me. But then to say he had discovered (not tested) a "secret AIDS treatment" was a little beyond the pale. Now, I don't have any trouble with somebody trying to improve my findings:

> Hoping to replicate the results in a placebo-controlled trial, Cummins sent
> Koech interferon tablets in January 1990. But Koech decided the results were
> so promising that it would be unethical to withhold the drug from a placebo
> population.

I guess it shouldn't come as much of a surprise that things didn't get any better from that point. I could completely understand the human desire not to have a placebo group with such a terrible disease.

***

I visited the American Foundation for AIDS Research (AmFAR) in the spring of 1990 and met Terry Beirn. We decided to test KEMRON® independently of the Africans, and Terry Beirn recommended Toronto, Canada, as a suitable test location. Terry wanted to demonstrate that clinical research could be conducted at the Community Research Initiative–Toronto (CRIT). With Terry's encouragement, some financial support from AmFAR, a cooperative Health Protection Branch of the Canadian Ministry of Health, and enthusiastic AIDS physicians in Toronto, we planned a clinical trial in Canada.

In July 1990, I flew to Tokyo and was at a meeting in which Secretary General Nakajima of the WHO met with representatives of HBL and the Japanese Ministry of Health and Welfare (MHW). Mr. Nakajima demanded that the MHW deny HBL export approval for any more testing of their human interferon in AIDS patients.

I was shocked that the top man at WHO would take such a position.

Not surprisingly, the Japanese MHW and HBL wanted to honor Secretary General Nakajima's request. The immediate impact of Nakajima's action was that HBL's human interferon could not be obtained for our clinical trial in Toronto.

After months of frustrating delay, Terry Beirn of AmFAR sent a fax to the WHO on Senator Ted Kennedy's letterhead and demanded an explanation why they were blocking the CRIT study. The WHO immediately replied by denying any involvement in blocking of the CRIT study. Given

Senator Kennedy's support of public health measures, the letter seemed to shake things loose.

In a few days, clinical supplies were released from Japan, and the CRIT study began in early 1991. I wrote to Senator Kennedy thanking him for his letter to the WHO. Terry Beirn called just as I was about to mail the letter. When I told Terry, he urged me not to send it to Kennedy. When I asked why, he said because the senator did not know anything about it.

Terry had swiped some stationery from the senator's office and written the letter on his letterhead and signed Kennedy's name! I could only think of how Ron Woodroof of the *Dallas Buyer's Club* would do anything possible to give people hope.

I can't say I'd ever take such a risk, but I understood why some people did. They have my respect, if not my approval.

Terry was an energetic, innovative thinker, and it was a great loss when he died on July 16, 1991. I'm sure God had a good laugh when He greeted Terry at the pearly gates and said, "Now, Terry, tell me about the incident with the Kennedy letter."

If I were God, I would've given Terry his angel wings the first moment I saw him.

The Community Program for Clinical Research in AIDS (CPCRA) was later named the Terry Beirn CPCRA to honor his pioneering efforts to generate research support for AIDS trials.

One hundred and fifty men participated in the Canadian study. For eight weeks, these men took either placebo or 50 or 100 IU of HBL's human interferon daily.

None of the benefits reported in Africa occurred in the population treated in Canada, and although it was disappointing, we reported our results. As we wrote in the summary:

> One hundred forty-nine patients of private physicians in Toronto, Canada, who were positive for human immunodeficiency virus (HIV), medically stable, and had CD4 cell counts of <700 cells/mm3 participated in a randomized, double-blind trial of placebo versus low-dose (50 U) versus high-dose (100 U) oral interferon-a. . . . Patients were observed at 4 and 8 weeks for assessment of adverse events and several measures of disease status, including CD4 cell count, B2-microglobulin, weight, and Karnofsy score. We detected neither short-term benefits nor adverse effects from oral interferon-a therapy.[117]

We immediately began to question the dose and timing of administration of the oral human interferon and the influences of race, gender, opportunistic infections, and diet on the different studies.

Why would black Africans respond differently from white Caucasians in Canada? It would be a question that would confound me for some time before I finally ran across a possible answer.

But by the time I did, would anybody still be interested in human interferon?

***

While the Canadian trial was in the planning stages, I received a call from Brendon O'Regan in San Francisco representing an organization called the Institute of Noetic Sciences. He asked if he could test low-dose oral human interferon in cancer patients in Nuremberg, Germany, with Professor Gallmeier. Arrangements were made to deliver HBL human interferon to Germany with the expectation that cancer patients would be treated.

A planning meeting was held in Germany with the investigators, people from HBL, me, and Steve Whitney of Mitsubishi Corp (HBL friends). I was impressed that Steve Whitney (from Oklahoma) simultaneously translated Japanese, German, and English in the meeting. I'm barely able to communicate in English in West Texas.

However, because of the excellent results reported out of Kenya about oral human interferon and AIDS, Professor Gallmeier and his colleague, Dr. Kaiser, decided to use the HBL human interferon lozenges on AIDS patients, instead of cancer patients.

Thirty German patients were studied for twelve weeks, six weeks on 200 IU human interferon once daily and six weeks on placebo. The Germans did not respond as the Kenyan population was reported to respond. Only a short time (at two and four weeks, but not six weeks) improvement in CD4+ cell counts was noted:

> There was only a slight transient increase in mean CD4+ lymphocyte counts after 4 weeks of treatment with natural human interferon, compared with a slight decline when placebo was administered. This effect reached statistical significance in a subgroup of patients only and was not sustained after 6 weeks. There were no significant changes in weight and clinical symptoms. All patients remained HIV-1-antibody-positive. Treatment related adverse reactions were not observed.[118]

We wondered again why a black African population responded differently from white Caucasians in Germany or in Canada.

A physician in New York State called to tell me how to help oral human interferon work in HIV-positive patients. "The patients need to eat kale," she said.

I called Dr. Obel in Kenya to ask about kale consumption in his population. He told me that kale was a staple in the diet of Kenyans. This is what the Mayo Clinic has written about kale:

> Kale is a nutrition superstar due to the amounts of vitamin K, B6 and C, calcium, potassium, copper and manganese it contains. One cup of raw kale has just 33 calories and only 7 grams of carbohydrates. So, it's a very diabetes-friendly/weight-friendly vegetable. Kale is a member of the cruciferous vegetable family along with cauliflower, Brussel sprouts [sic], cabbage, broccoli, collard greens, kohlrabi, rutabaga, turnips, and bok choy.[119]

Was kale the missing factor in whether human interferon would work at maximum capacity? Maybe it's something in the cruciferous vegetable family that provides the proper blend of nutrients that allows human interferon to work at maximum capacity.

I must confess I felt blindsided by this newest addition to the interferon story.

There were so many questions I had that I couldn't answer. Were animals responding more positively to human interferon because they had a better nutritional status than the AIDS patients? In my animal studies, the animals all received the same diet.

Was kale needed in all conditions, or just AIDS?

Could it be any member of the cruciferous vegetable family?

I'm not sure whether I should have started talking about kale at this time, but I chose not to do it. I felt like so many things were happening I didn't want to bring up the kale issue.

***

Other AIDS studies using oral human interferon were conducted from 1989 to 1994 in Kenya, Zambia, Uganda, the Philippines, Thailand, New York, Puerto Rico, Japan, and Canada involving over a thousand patients.

A study at Mahidol University in Bangkok tested placebo and 100 or 200 IU of HBL oral human interferon given daily for six months. Dr. Prasert

Thongcharoen, who was the principal investigator in Thailand, reported a significant weight gain and relief of symptoms in AIDS patients given 200 IU per day:

> At 24 weeks, patients given 200 IU HBL human interferon had a significant (P<0.05) weight gain, as compared to patients given placebo or 100 IU IFN-a. The patients given 200 IU HBL human interferon also had a significant (P<0.05) improvement in the clinical complaints of fatigue/weakness and appetite loss/weight loss compared to patients given placebo. In addition, patients given 200 IU HBL human interferon had trends toward greater improvement of overall clinical complaints and greater self-evaluation scores. No significant differences were observed between the treatment groups with respect to the other variables evaluated, including CD4+ cell count. Adverse reactions were not reported by any patient. Eleven patients died during the study due to HIV-related opportunistic infections.[120]

There are a couple of observations about the Thailand study. How sick were these patients at the beginning of treatment? Sixteen percent of the patients died in less than six months!

I'd say they were looking at a pretty sick population. Even with more than 16 percent of the patients dying, there were positive results.

The group treated with the highest daily dose, 200 IU of HBL human interferon, recorded an average weight gain of 2.2 kg, or just a little under five pounds.

And the lack of a positive response in the CD4+ cells? Maybe their CD4+ cell counts were so low to begin with that it would take further treatment in order to improve those levels. When one looks at the dramatic drop in patient reports of fatigue/weakness, most pronounced in the 200 IU group, but also significant in the 100 IU group, it becomes clear that a definite improvement was being seen in patients using human interferon.

In a study with Dr. M. Mukunyandela of the Tropical Diseases Research Center in Ndola, Zambia, a total of 147 HIV-positive patients were treated for twenty-four weeks with lozenges containing 150 IU of human interferon or a matching placebo. One group took active lozenges one week and placebo lozenges the next week. One group took oral human interferon daily, and one group took placebo daily. The treatment group that took 150 IU human interferon lozenges continually experienced significantly fewer new HIV-related opportunistic infections, compared to the placebo group, while both groups of human interferon-treated patients tended to gain CD4+ cells, while the CD4+ cells of the placebo group remained flat:

> Patients given 150 IU HBL interferon continuously (150 IU cont.) had no increase from baseline in the mean number of HIV-related signs and infections per patient while both the placebo group and the group given 150 IU HBL interferon intermittently (150 IU int.) had mean increases during the study (P=0.001). Both HBL interferon treated groups had significantly greater (P=0.05) resolution of HIV-related skin infections present at baseline compared to the placebo group. In the 150 IU cont. group only 1 of 44 patients (2%) without mucocutaneous herpes simplex virus infection at baseline developed such an infection during the study as compared to 6 of 47 (13%) in the 150 IU int. Group and 12 of 48 (25%) in the placebo group (p<0.006). In addition, patients in the 150 IU int. group had significant trends toward a greater mean reduction in global symptom score (P<0.025) and small increases in both mean absolute CD4+ count (P<0.05) and mean CD4% (P<0.001) compared to both groups.
>
> No serious adverse experiences were reported in this study. Sixteen of 147 patients died due to opportunistic infections during this study. No difference in mortality rate was noted between the groups and none of the deaths were judged to be related to HBL human interferon.
>
> In conclusion, in this study in Zambia, the administration of a low dose of a natural human interferon to 147 symptomatic, HIV-1 infected patients for 6 months was safe and led to improvements in clinical signs and symptoms of HIV disease and to modest improvements in CD4+ counts.[121]

Again, I must note how sick of a population was being studied. Across all three groups over a twenty-four-week period, there was a death rate of just above 10 percent!

Besides soldiers fighting in a war or senior citizens in nursing homes during COVID-19, what other group would have such a high fatality rate in a time period of less than six months? Let's go over some of the positive results from this study.

The group that got 150 IU of oral human interferon (still an amazingly low dosage compared to many of the studies) on a daily schedule only developed a herpes simplex infection in 2 percent of the patients.

The group that got 150 IU of oral human interferon for one week, then one week with a placebo, developed a herpes simplex infection in 13 percent of the patients.

The placebo group that never got oral human interferon developed a herpes simplex infection in 25 percent of patients.

In addition, both groups that received oral human interferon showed a significant reduction in skin conditions. The group that consistently received oral human interferon also had the greatest reduction in their global system score.

***

From these many studies, data accumulated that allowed us to choose an oral human interferon dose for the NIH clinical trial (DATRI 022), which started in 1990. A total of 560 AIDS patients were to be enrolled to take three different sources of oral human interferon versus placebo.

After years of planning, it was announced in February 1995 by the AIDS Research Advisory Committee that the oral human interferon AIDS study was approved. In July 1995, the NIH ordered the clinical supplies for delivery to enroll patients. Unfortunately, because of slow enrollment, the study was halted in June 1998, and no meaningful results were possible.

Because 560 patients were needed to provide sufficient "power" to analyze the data, we objected to the publication, which drew conclusions based on too few patients. Here is what was reported in the summary:

> Although the trial was designed to enroll 560 study subjects and was prematurely terminated because of the slow accrual and discontinuation of participants, the small difference among the arms in the primary and secondary endpoints do not support claims of efficacy for the measures studied.[122]

Do you understand this study, which was supposed to have 560 patients in several different groups in order to have enough power to give a clear answer, failed because not enough people participated?

The DATRI 022 data have been reanalyzed, and we reached a different conclusion from the one published in the *Journal of Acquired Immune Deficiency Syndromes* (*JAIDS*).

The summary in *JAIDS* concluded by adding "the small differences among the arms in the primary and secondary endpoints do not support claims of efficacy for the measures studied."[123] We disagree with this conclusion for two main reasons:

1. Based on the power calculations for the trial, there were too few
    subjects enrolled and far too few subjects who finished twenty-
    four weeks of treatment to draw any definitive conclusions on
    treatment efficacy.
2. In those who did finish twenty-four weeks, significant differences
    were noted between treatment groups in CD4+ cell counts,
    depending on how the data are examined. There are additional
    areas of concern with the data, as detailed below.

The DATRI 022 study was designed to enroll five hundred and sixty
subjects to detect differences in treatments. It was reported in *JAIDS* that
only 247 evaluable study subjects were enrolled and data on CD4+ cell
counts of eighty-nine subjects who completed twenty-four weeks were
presented.

Therefore, 44.1 percent of the necessary number of subjects was enrolled
and only *15.9 percent* of the projected 560 subjects completed the study.

The *JAIDS* conclusion that these data "do not support claims of efficacy" for the measures studied therefore is not correct. The protocol design
was such that a larger number of subjects was needed to detect a significant
difference between treatment groups.

It is a statistical truism that one cannot draw a definitive conclusion
based on an underpowered clinical trial. That should have been acknowledged in the *JAIDS* publication and should have tempered their conclusions.

Due to the inadequate number of study subjects, definitive conclusions
should not have been drawn.

***

Lozenges containing 150 IU of HBL human interferon were delivered to
the WHO in 1991 so WHO could conduct a clinical trial of AIDS patients
in Uganda.[124] A total of 559 patients cooperated in the trial and were given
either placebo or human interferon daily, orally, for up to six months.

However, most of the patients were treated for thirty days or less.

Many patients died during the study due to "diarrhea," tuberculosis,
and other complications associated with AIDS. Many patients had no or
very few CD4+ cells (mean baseline CD4+ cell count was only 60 cells/
cu.mm, whereas >800 cells/cu.mm are normal) upon entry to the study,
thereby forecasting a probable early death.

Indeed, nearly a third of the patients died during the study.

Even under the severe circumstances of the WHO study, a CD4+ cell count improvement was noted for eighteen weeks.

But the HBL oral human interferon therapy did not provide a beneficial effect on mortality.

For months, WHO was asked to share the data so we could determine if there was a subset of patients (e.g., males with initial CD4+ cell counts >200) who might have benefited from treatment with oral human interferon.

WHO did not provide the data for analysis, even though we asked several times.

We now believe that oral human interferon benefits most readily occur in patients with baseline CD4+ cell counts >200, treated for six months.

There is probably no single agent that can be given to AIDS patients with a CD4+ cell count below sixty that will seem beneficial.

Many of the patients in the WHO study entered with CD4+ cell counts of *zero*. It is not surprising that nearly a third of the patients died. Why would WHO enroll anyone with a CD4+cell count of zero? Was this the revenge of Secretary General Nakajima because of Senator Kennedy's letter?

Does this action by the WHO look fair?

***

The Nation of Islam started selling oral human interferon in the United States as a cure for AIDS.

The FDA ignored the activities of the Nation of Islam and the sale of oral human interferon by other organizations unrelated to the Nation of Islam. ABI complained to the FDA about the openly defiant sale of oral human interferon and its advertising in the United States. The controversy with the Nation of Islam was summarized by *Newsweek* in January 1993:

> Two years ago, a Kenyan researcher named Davey Koech shocked the world with an announcement about the treatment of AIDS. He claimed that when his patients chewed wafers laced with tiny amounts of alpha interferon, a drug used at vastly higher doses to treat some cancers, their symptoms vanished, their immune systems rebounded and some cleared HIV from their blood. The Kenyan government embraced the report as an epochal triumph, and U.S. black media were soon sharing in the jubilation.[125]

Newsweek was kind enough to publish a picture of me in the article standing in front of cattle. And *Newsweek* also did a good job of making certain my contribution to the issue was highlighted in the article:

> Though championed as an African innovation, low-dose oral interferon is the brainchild of a white Texan named Joseph Cummins, who conceived it as a treatment for respiratory diseases in cattle. Interferons are natural chemicals that cells use in minute quantities to fend off various assaults. Doctors sometimes administer them at thousands of times their natural concentrations to fight viruses or stimulate the immune system. But Cummins has used low oral doses to treat both human and animal illnesses, and the results have been intriguing.[126]

I was generally pleased at how *Newsweek* framed the issue. I hope you realize that what I've been saying is in line with news accounts of the time. Specifically, that the idea of low-dose interferon was my contribution and that interferon was found in the body in minute quantities during an infection. Science acknowledged this fact, but their answer was to "administer them at thousands of times their natural concentrations to fight viruses or stimulate the immune system."

If we know that minute quantities of interferon are enough to clear infections and pathogens, how in God's name do you decide it's a good idea to give it at "thousands of times" what you would find in a healthy organism?

The Kemron story went on for years, and it just got crazier.

The *Chicago Tribune* also published a front-page article on the Kemron controversy and the Nation of Islam selling the "African" cure in March 1995:

> In the past two years, the U.S. government has poured $571,521 into a sparsely furnished clinic in a low-income District of Columbia housing complex, where the Nation of Islam's chief doctor sells a "miracle drug" he claims will cure AIDS.[127]

Again, let's go down the chain of events. I tried to work with Kenyan officials, and when it came back to America, it's supposedly an "African remedy." And I had no idea who was preparing it and under which conditions it was being prepared, and this Kemron gets picked up by the Nation of Islam as an AIDS cure. Was I controlling any of this? No.

"If it is promoted as a cure, that's just criminal," said Dr. Wilbert Jordan,
director of the AIDS program at King/Drew Medical Center in Los Angeles,
who has used interferon in combination with other medications.

Some studies indicate that the pills helped some AIDS patients gain
weight, but other tests failed to prove even those limited benefits.[128]

Did I want anyone selling some dubious knockoff of my product to the
Nation of Islam? No, I didn't. Every part of me found that situation to be
revolting.

There's no room for ego when people are suffering and dying. And
whatever one might think of the Nation of Islam, I didn't want them getting an inferior product, so I actively tried to get them my own formulation.

I wrote twice to the FDA, asking if I could give away free oral human
interferon to patients. That way, the patients would not be forced to go
to great lengths to purchase "interferon" of unknown content, unproven
effects, and questionable quality.

I was told that I would *not* be allowed to provide free human interferon.

Open sales of oral "interferon" slowed to a trickle after the NIH published their report that the treatment was useless.

Sometimes I wonder if we have the wisdom to survive as a species.

***

In the summer of 1990, Jim Corti came from Los Angeles to Amarillo. He
wanted to hear firsthand about our experiences with oral human interferon.
Jim was a prominent member of Search Alliance, an AIDS support group
in Southern California. Jim traveled the world seeking treatment for AIDS
patients. He is one of the heroes of a book written by Jonathan Kwitney
titled *Acceptable Risks*, published in 1992.

Jim became interested enough in oral human interferon that he acquired
some liquid human interferon and enlisted the cooperation of physicians
who were members of Search Alliance. The physicians set up a quick study
to test their oral human interferon in AIDS patients but quickly became
disillusioned with the results they saw. Within a few weeks, all the physicians abandoned the study, except Dr. Wilbert Jordan.

Even though low-dose oral human interferontherapy for AIDS patients
was severely criticized by the AIDS "establishment," Dr. Wilbert Jordan
of Los Angeles courageously and steadfastly used oral human interferon
with beneficial effects, as he reported in the *Journal of the National Medical*

Association. This is what he reported in the discussion section of his article in 1994:

> All of these studies were initiated because the author's patients, after hearing the widely publicized, remarkable results reported by Koech et al., decided to begin therapy with oral [natural human interferon]. The first study was designed to test whether the addition of AZT to natural human interferon therapy might cause even greater increases in CD4+ cell count than anticipated, based on the results obtained at KEMRI, for [natural human interferon] alone.
>
> Although dramatic increases in CD4+ cell count were not seen in this study, 75% or greater of the patients in both groups exhibited modest mean increases in their CD4+ cell counts. Overall, 252 patients with a baseline CD4+ cell counts of 356 cells/mm3 were given natural human interferon for 1 year and experienced a mean increase of 76 cells/mm3, a 21% increase in mean CD4+ cell count. These data suggest that natural human interferon may lead to moderate increases in numbers of CD4+ cells in HIV-infected patients when given in low doses by the oral mucosal route for up to 1 year.[129]

As a side note, I must add that these results were obtained without knowledge of the patients' consumption of kale or other cruciferous vegetables, which might account for the remarkable results obtained in the Kenyan study.

Let's talk about how remarkable these results were, given what was happening in 1994. The CD4+ levels were the best indication of the progression of the disease. If they remained stable, one could generally expect to go on living. These results showed a 21 percent average increase in CD4+ levels, meaning these patients were moving away from death and toward recovery. Now, was it taking them all the way to recovery? No. But it provided time to figure out if other therapies, in combination with human interferon, might lead to a recovery.

Dr. Jordan tested HBL oral human interferon under a physician's "IND" (Investigational New Drug Application). Dr. Jordan has given the oral human interferon technology his support and has demonstrated his pioneering research spirit. Against all the critics who, without testing oral human interferon, deny the effects of oral human interferon, Dr. Jordan offers his years of clinical observations of its benefit.

He reported, but did not publish, his observation that HIV-positive men given oral bovine interferon did not develop anal warts. Papillomaviruses in the cervix and in the mouth seem to be treatable with oral human interferon.

***

In 2018, I published a short commentary in the journal *Current Research on HIV/AIDS* on the studies that had been done regarding human interferon and HIV/AIDS. In the summary section, I wrote:

> Orally administered human interferon has been tested in 23 studies of patients infected by HIV-1. The preponderance of data (review published in 2004 in AIDS Vaccines and Related Topics, pages 11-30) suggests that orally administered human interferon is useful in the management of opportunistic infections in HIV+ patients. AIDS is an epidemic out of control in some parts of the world. The clinical efficacy observed with oral human interferon in HIV+ patients and the ease of administration, lack of toxicity, room temperature stability and low cost indicates that oral human interferon may have a role in helping manage AIDS.[130]

A double-blinded, placebo-controlled AIDS study was started in early 1989 in Amarillo with the cooperation of the local health department and Texas Tech Health Sciences Center. HIV-1 seropositive patients were given placebo or oral human interferon at 0.1, 1 or 10 IU/lb of body weight daily for seven days.

They then skipped a week and repeated the treatment.

The average time on "every other week" treatment of 171 patients in the study was nineteen months.

The middle dose of human interferon was shown to provide a clinical benefit. However, compliance was extremely poor, and not much useful information could be gained, even though the data were published in *Biotherapy* in 1998.[131] Funding for this clinical study was granted by Mesa Petroleum in Amarillo, then operated by T. Boone Pickens. I wish we'd been able to perform a more comprehensive study, as I believe it would have shown strong clinical benefits.

The proper research and protocols must be used in testing low-dose interferon, and if they are, I believe the science will clearly show an enormous benefit from this medication.

CHAPTER 9

The State of Interferon Today

Veterinarians can purchase low-dose human interferon from a compounding pharmacy but should ask for human interferon assay procedures and human interferon titers in the lot purchased. Veterinarians can purchase concentrated human interferon (e.g. 5 million IU/vial) and dilute the concentrated human interferon to the desired oral dose.

Such diluted human interferon is stable for only three to four days at room temperature, so most vets have probably been giving "placebo" instead of active human interferon. However, human interferon may be stable up to six months, if refrigerated. The addition of 0.5 percent bovine or human serum albumin will help stabilize the diluted human interferon for many months. It is remarkable to me that a product that has been safely used in animals for decades is unavailable to the human population.

The dose identified in animal studies in which human interferon has been tested is a very low concentration. We are so close to a golden age of health. We only need to take a few more steps to that victory. Nature has offered us the tools.

We need only have the wisdom to accept her gifts.

***

Was it simply a blind spot that made scientists vigorously pursue high-dose interferon treatment, or was it part of a plan by the pharmaceutical companies to cripple this promising therapy?

Because I wasn't there in the room with those researchers or in the corporate boardrooms, I can't give you a definitive answer. But I have my suspicions. Maybe it's because I'm an older man, but I've generally found that if people are honestly looking for an answer, they find it.

History records that pharmaceutical companies developed high-dose interferon to treat certain cancers, hepatitis B & C, and warts. Big Pharma companies in the United States, Europe, and Japan made many millions of dollars selling high-dose interferon before the toxicity and other factors led to market withdrawal. The use of high-dose injected interferon as a medical treatment defies explanation.

A couple of glasses of water a day is great.

Drinking the amount of water in a swimming pool in a day will kill you.

I want to make sure you understand the enormous difference between a nanogram and a milligram. Let's say you're having trouble sleeping and you think that maybe you'll take a sleeping pill this one time. You don't want to make it a habit, but it's really been a tough day, and that one sleeping pill is likely to get you back on track.

In this example, that one sleeping pill is equivalent to one nanogram.

One milligram is equivalent to a million sleeping pills. Yes, that's right. A milligram is one million times more than a nanogram. I hope you now understand why, when researchers start talking about using interferon in milligram doses, it drives me insane.

When drugs such as insulin and erythropoietin were developed by Big Pharma, these drugs were given to patients at doses comparable to normal (endogenous) concentrations. Insulin and erythropoietin would be quite toxic, probably fatal, if injected into patients at a hundred or a thousand times the normal concentration.

Many millions of dollars in profits were made by Big Pharma selling high-dose human interferon for hepatitis. High-dose human interferon is coming off the market because newer, safer, and more effective drugs have largely replaced it.

I believe high-dose injected human interferon has no place in medicine.

***

A cubic foot of air may be teeming with up to 100 million microbes.[132] The air we breathe is loaded with pollen, fungi, mold, viruses, bacteria, chlamydia, and allergens to which we may react. Trillions of microorganisms live in and on us. Because of the large volume of the passengers, less than half the DNA we carry is human. We are truly a "bed and breakfast" for the trillions of microbes in our gut, in our respiratory tract, and on our skin.

How do we survive the onslaught from trillions of microbes and allergens? Not only do we survive, we usually remain asymptomatic because of our miraculous immune system. However, over a lifetime, autoimmune diseases such as lupus, diabetes, multiple sclerosis, chronic obstructive pulmonary disease, idiopathic pulmonary fibrosis, Behçet's disease, and rheumatoid arthritis can result from the chronic inflammation induced by the immune system.

The cost of the goods to produce enough interferon for a single low-dose application is less than a dime. I sincerely believe that a clear understanding of interferons may be the single biggest breakthrough in all human health.

***

Importantly, human interferon is not a species-specific signaling molecule. Interferon from one species will also work in a different species. For example, cow interferon alpha works in cows and humans. Human interferon seems to work the best, though, in all species. Human interferon can be safely administered orally to cattle, cats, dogs, horses, and pigs.

Oral human interferon induces many beneficial effects in cattle, including increased resistance to infectious diseases, weight gain, and morbidity benefits in feedlots. In cats, human interferon (10 IU/kg or much less) exhibits statistically significant disease-sparing capabilities when given orally. Oral human interferon in dogs is helpful in treatment of parvovirus and KCS.

Oral human interferon in pigs helped manage TGE, rotavirus, Streptococcus suis, and PRRS. Oral human interferon in horses reportedly helps manage IAD.

The first part of this book talked about my history and summarized many different studies performed in companion animals and livestock on the benefits of oral human interferon therapy. From these, a consensus has emerged.

Oral administration of human interferon is effective, but at surprisingly low doses when compared to doses of human interferon designed for parenteral administration in humans. Unwanted side effects of orally administered human interferon are rare. The optimal dose usually appears to be less than 1.0 IU/kg; this can be administered directly into the oral cavity or can be incorporated into the drinking water.

How does such a trivial oral dose of human interferon have such dramatic systemic effects? The answer lies in our emerging understanding of the enteric microbiome and the emerging consensus that this interaction programs systemic protective immunity.

Receptors for human interferon have been identified in the oropharynx, and these receptors, when engaged with human interferon, modify expression of genes activating diverse effects including antigen processing and presentation, leukocyte migration, lymphocyte activation, immune effector and modulation functions, apoptosis, and hematopoiesis.

The activation signals from the microbiota and from oral human interferon are transmitted systemically to bone marrow progenitor cells and T and B cells residing in the organized lymphoid tissues associated with the enteric and oral mucosa. A cascade effect of "resistance" is generated that is characterized by induction of IL-2 production, immunoglobulin production, and activation of regulatory and cytotoxic T cells.

When viewed in this context, local human interferon action in the oral cavity is rapidly translated into systemic effects. Moreover, these beneficial protective effects may be transitory, necessitating repeated oral administration of the human interferon trigger.

Finally, like all signaling molecules, there exist strong dose-dependent down-regulating feedback loops wherein high doses appear to saturate these receptors, preventing the translational movement or cross-linking of these receptors on cell membranes, thereby negating the positive cellular response signal.

While details still need to be delineated in this oral-to-systemic pathway, the outlines of it are clear and consistent with what we know about other cytokines and cell-signaling molecules. They are also consistent with the observed beneficial effects of oral human interferon in domestic animals. We would expect to see a similar beneficial effect in humans.

***

Dr. Ericsson of Baylor University became interested in oral human interferon in 1989. As a neurologist, he used oral human interferon to treat polymyositis, multiple sclerosis (MS), Parkinson's disease, and other difficult clinical conditions. His successes in the use of oral human interferon are detailed in numerous publications.

Polymyositis is an inflammatory muscle condition that causes degenerative changes in muscles, causing weakness, pain, and muscle atrophy. Other symptoms can include fatigue, fever, weight loss, and pain or tenderness in muscles and joints. Patients will often present with infections of the Epstein-Barr virus, herpes, and cytomegalovirus, all of which are DNA viruses. Yes, you might say these patients aren't just infected with one virus. It seems an entire asylum of viruses somehow got opened up inside them. In his study of twenty-five patients treated with interferon, Ericsson wrote:

> These patients demonstrated not only dramatic clinical improvement, but a significant reduction of viral titers over an 8-week period. Moreover, each of these patients have been followed for a minimum of 90 days following the conclusion of the study with continued clinical improvement. It appears from the data that oral alfa interferon is effective not only for the treatment of DNA viral induced polymyositis syndrome (chronic fatigue) but dramatically reduces the viral titers associated with the altered disease state. Furthermore, as noted by the clinical lack of toxicity, the drug is remarkably well tolerated without significant severe biological reactions.[133]

I will agree that there are several things that are unclear to us, such as the contribution of each virus to the disease. However, the continued persistence of these viruses lets us know that in some way, these viruses, either on their own or in concert, are unbalancing the immune system. But interferon is clearly lowering viral titers, and that is allowing the body to heal.

Dr. Ericsson published about clinical studies of various neuromuscular diseases, such as multiple sclerosis (MS) and Parkinson's disease, treated with low-dose oral human interferon at 1500-3000 IU/dose once daily. In his research on MS, he followed fifteen patients with the disease, four males and eleven females, ranging in age from thirty-five to sixty-three years old. This is what he reported:

> There was never a sudden dramatic change in any of the patients, but the following observations were made. Aside from a feeling of well-being and increased appetite (50%), little change was noted for the first two weeks after initiating therapy. About 2–3 weeks after therapy all patients were capable of activities that they had not been doing for months prior to the study. For example, the ability to sign checks, write legibly, or walk with greater ease was apparent. This improvement has continued but was not uniformly associated with a change in the neurological status (note change in neurological status

+1.07), and several patients are now walking who were wheelchair dependent for over 6 months prior to alfa interferon treatment.[134]

If you know anybody who suffers from MS, you understand how debilitating a condition it can be. Many people can go for years without significant symptoms of MS, only to have a flare that lasts for weeks, months, or even years. The simple daily activities of being able to legibly write your name on a check or sign a credit card receipt or even take a walk can be impacted.

Another terrifying medical condition without a clear cause is Parkinson's disease. However, we believed there might be a significant immune system component to the somewhat unusual range of symptoms experienced by Parkinson's patients. As we wrote in the opening of the Parkinson's section:

> The inclusion of Parkinson's disease in this group of alfa interferon treated patients may at first seem bizarre. In light of the "on/off" effect, the known effect of sudden ability to move and function with a loss of rigidity and tremor followed by dramatic loss of this effect (often occurring within seconds), and post-synaptic receptor co-dopaminergic blockade, the immune system may not be entirely faultless in the myriad of neurological symptoms of Parkinsonism.[135]

Clinicians who work with Parkinson's patients report it is one of the most challenging and heartbreaking diseases. Not only does it seem to be striking people at a much younger age, but in the past where it was observed that most of the problems were in the motor system, there appears to be an increasing rate of cognitive problems with the disease.

In the study, we had seventeen patients who were tracked for twelve weeks. Nine were male, eight were female, their ages ranged from forty-nine to seventy-two years, and their mean disease duration was 12.3 years:

> Each patient kept an hour diary and plotted the mean effective hours/day of activity for the week prior to initiation of alfa interferon therapy and each week thereafter for the duration of the study. Overall, there was an increase in the useful hours in each day by 2.5 hours/day. The most dramatic increase was in a patient who had 1.7 hours of activity before treatment that increased to over 7.5 hours of activity after interferon therapy . . .
>
> These data suggest that the immune mechanism is not passive in Parkinsonian patients on chronic administration of 4-DOPA/carbidopa.

> Furthermore, alfa interferon is well-tolerated, and the effects of treatment
> warrant a continuation of this study.[136]

In three months, we were able to add an average 2.5 hours a day of activity to this population of Parkinson's patients. And this was with an intervention that was well tolerated. When we fully understand how much the malfunctioning of our immune system (possibly due to pollutants or pathogens) is contributing to many diseases, I think we will be shocked.

Dr. Ericsson also demonstrated courage and scientific curiosity in testing low-dose oral human interferon and pairing it with an antiviral drug (ECFD-HIV with AIDS patients in West Africa). Using interferon in combination with another medication to improve results has been apparent to me for some time. That synergistic force might yield some truly remarkable results, as was reported in this study and hinted at in others:

> A small-group pilot double-blind placebo-controlled study was conducted by
> the Buganda Medical Institute in Tanzania using VIRON (alfa interferon
> combined with ECFD-HIV) in which 12 patients were treated with the active
> drug while 11 patients were assigned placebo. Over a 60-day period, these
> HIV positive patients with AIDS treated with VIRON had a mean increase
> in their CD4 counts of 605, while those receiving placebo had a decrease of
> 102.[137]

Let's be clear on what this means. If your CD4 cells are increasing rather than decreasing, that means you are probably recovering, not dying! This was yet another example of encouraging results not being followed up with larger, well-designed studies.

***

Dr. Brod is a pioneer in the use of low-dose oral human interferon in MS, diabetes, and other conditions. He called his method of delivery "ingested" human interferon because he placed the human interferon in the mouth and within seconds washed it into the stomach with at least five ounces of water. Interferon within the oral cavity reacts with interferon receptors within milliseconds, so oral activity probably occurs before interferon is ingested.

In January 2003, Brod published what he called a unifying theory of interferon usage for autoimmune conditions, based on his extensive research:

We have proposed a unifying hypothesis of the etiopathogenesis of autoimmunity that defines autoimmunity as a type I [interferon] immunodeficiency syndrome. We have examined toxicity and potential efficacy in three phases (type 1 diabetes, rheumatoid arthritis, multiple sclerosis) and one phase II clinical trials in multiple sclerosis (MS). In a phase I open-label trial in type 1 diabetes, ingested [human interferon] preserved residual beta cell function in recent onset patients. In a second phase I trial, treatment of rheumatoid arthritis (RA) with ingested [human interferon] reduced the secretion of interleukin-1 (IL-1), a proinflammatory cytokine.

In a third phase 1 trial in MS, there was a significant decrease in peripheral blood mononuclear cell (PBMC) IL-2 and IFN gamma after ingesting [human interferon]. . . . These studies suggest that [human interferon] may have a potential role in the treatment of autoimmunity.[138]

I see a boldness in Dr. Brod's claim, putting his reputation as a researcher at the University of Texas on the line. I was making specific claims in various animal and human diseases after testing. But Brod was saying that what these diseases had in common was an underlying autoimmunity, and that interferon could treat it.

After various successful trials in different conditions, one needs to ask if there's some underlying mechanism at work that enables someone to do this. In order to be most successful, that person should ideally have a great amount of respect in the community and be willing to place that prestige at risk. Brod had that capability, was a good ally to have in this fight, and was willing to move forward with exceptional courage.

***

Dr. Gerald Holman of the Hospice at St. Anthony's (Amarillo) tried Roferon (FDA approved human interferon) in low oral dosage in five cachectic cancer patients. A few patients seemed to experience notable appetite stimulation, similar to reports in veterinary medicine. In my opinion, Dr. Holman did not use the "best stuff" (i.e., natural human interferon) but instead used genetically engineered human interferon. He used a dose that was probably not optimal when he treated terminal patients for five days.

Yet even under these imposing conditions, Dr. Holman reported that "There appeared to be some benefit in appetite and general well-being in three of the five patients, and possibly to a small degree in one other . . ."

Dr. Holman's observations led to a study at the Don & Sybil Harrington Cancer Center in Amarillo, where placebo or genetically engineered human interferon at 0.1, 1.0, or 10 IU/lb of body weight was given to fifty-four "dying" cancer patients experiencing weight loss (at least 5 percent) or anorexia. I don't believe he ever published his results but detailed them in a personal letter to me.

These privately reported findings led to my own research into the question that was published in the *Journal of Interferon and Cytokine Research* in 1999:

> In a double-blind placebo-controlled trial, 57 adult subjects with disseminated malignancies were given orally low doses of recombinant human interferon-alpha at 0.05 IU, 0.5 IU, or 5.0 IU/kg body weight. The objective was to determine the efficacy of orally administered human interferon on appetite stimulation and/or weight loss prevention in anorectic cancer patients. Almost two-thirds (64%) of the subjects given 5.0 IU/kg reported an increase in appetite or body weight after 5 weeks in contrast to only 29% of the placebo treated subjects. . . . Additionally, the 5.0 IU/kg treated group experienced half as many deaths as the control group by the conclusion of this 91 day trial.[139]

The pattern that keeps repeating itself is an increase in appetite, weight gain, and a reduction of the death rate.

Based on that report, and observations in other human and animal trials, low-dose oral HBL human interferon was tested in cancer patients at Wayne State University, Detroit, to help control one common side effect (oral mucositis) of chemotherapy. A beneficial response was noted in most patients and submitted to the American Association for Cancer Research.[140]

In 1978, studies of human interferon in cancer began when human or bovine interferon was given orally to animals or humans with cancer. As a result of this effort, several patents were issued for the use of oral human interferon. US Patent No. 5,017,371 issued in 1991 is titled *Method for Reducing Side Effects of Cancer Therapy.* The independent claims are that human interferon in contact with the oral or pharyngeal mucosa reduced the side effects of radiation or chemotherapy in cancer patients. This claim was successfully tested, as noted above, at Wayne State University in head-and-neck cancer patients treated with 5-FU, otherwise known as fluorouracil, a common medication used to treat cancer.

US Patent No. 5,824,300 issued in 1998 is titled *Treatment of Neoplastic Disease with Oral Interferon.* The claims are made that human interferon in contact with the oral or pharyngeal mucosa is efficacious in treating malignant lymphoma, melanoma, mesothelioma, Burkitt's lymphoma, nasopharyngeal carcinoma, leukemia, and Hodgkin's disease.

Successful treatment of melanoma with natural bovine interferon was first reported in US Patent No. 4,462,985 issued in 1984 and later published in 1993 in an article I cowrote titled "How it Began."[141]

The issued US Patent in 1984 is titled *Delivery of Biologically Active Components of Heterologous Species Interferon Isolates.*

Despite low-dose oral human interferon's safety, inexpensive nature, and apparent aid in the successful human treatment of osteosarcoma, melanoma, renal cell carcinoma, prostate cancer, acute myelogenous leukemia, and appendiceal adenocarcinoma with signet ring cell, oncologists in the United States generally denigrate our research efforts. Our observations on cancer patients are anecdotal and easily dismissed.

However, successful oral human interferon studies in cats with feline leukemia and mice with various cancers are well controlled. Apparently, herding cats is much easier than herding humans. Others have also recognized the benefits of oral human interferon in the treatment of cancer.

MG Tovey in France has US Patents Nos. 5,997,858 and 6,660,258, issued in 1999 and 2003 with titles of *Stimulation of Host Defense Mechanisms Against Tumors* and *Oromucosal Interferon Therapy: Marked Antiviral and Antitumor Activity.* These patents and publications by Tovey and his colleagues clearly demonstrate the safety and efficacy of oral murine interferon in treating cancer in mice.

The mechanism of action of oral human interferon in cancer is complex, and I believe it includes enhancement of natural killer (NK) cell activity (NK cells are some of the first cytotoxic lymphocytes that go after viruses and tumor cells) and modification of expression of interferon susceptible genes.

***

After reviewing HBL's Drug Master File, ten IND applications were sent to the FDA to allow for testing. The FDA allowed the testing of oral HBL human interferon in US human clinical trials in fibromyalgia, Sjögren's syndrome, and HIV+ patients with oral warts.

My son, Marty Cummins, prepared the protocols, handled all the FDA paperwork, interacted with the study site personnel, negotiated with the principal investigators, and made it possible to test oral human interferon in humans in FDA-approved studies. (I must admit some paternal pride that my son is so smart.) Here are some study results.

*****

Sjögren's syndrome is a chronic autoimmune disorder distinguished by dryness of the eyes and mouth. It exists as a primary disorder or in association with other autoimmune diseases. Patients with primary Sjögren's syndrome generally have clinical signs such as rash or joint, lung or kidney inflammation. Typical signs and symptoms generally include the sensation of burning eyes; mouth, skin, nose, and vagina dryness; swallowing difficulty; throat pain; and fatigue. Although Sjögren's syndrome is not life-threatening, it can seriously impair quality of life.

The Sjögren's Syndrome Foundation, Inc., estimates that the disease affects approximately two to four million people in the United States. The incidence of Sjögren's syndrome worldwide is similar to its incidence in the United States. Women constitute 90 percent of Sjögren's syndrome patients.

Oral human interferon therapy helps relieve dryness associated with Sjögren's syndrome and improves secretory function.

At a cost of over six million dollars, two twenty-four-week Phase III clinical trials were conducted to test the oral use of HBL human interferon in the treatment of primary Sjögren's syndrome.

Results of both Phase III clinical trials demonstrated an improvement in saliva production in treated patients. The studies were double-blinded, placebo-controlled tests in which a total of 497 patients were treated orally three times daily for twenty-four weeks with either 150 IU of human interferon or placebo. This is what we reported:

> In human interferon-treated patients, increases in UWS [unstimulated whole saliva] flow correlated positively and significantly with improvements noted in 7 of 8 symptoms associated with oral and ocular dryness. The coprimary endpoints of stimulated whole saliva flow and oral dryness were not significantly improved in the human interferon group relative to placebo. No significant differences were found between the groups with respect to overall adverse event incidence or severity. CONCLUSION: human interferon given

at low dosage by the oromucosal route can significantly increase UWS flow
in patients with primary Sjogren's syndrome, without causing significant
adverse events.[142]

Analysis of participants who completed the trials, designated as evaluable patients, found a significant increase in UWS production among the
human interferon-treated patients, as compared to placebo. Increases in
saliva production are important to the Sjögren's syndrome patient, since
low saliva production is a problem that is present in over 90 percent of the
patients.

Importantly, in human interferon-treated subjects, there was a significant correlation between increases in saliva production and a decrease in the
symptoms of Sjögren's syndrome, including oral, throat, and nasal dryness,
and the ability to swallow foods.[143]

This research showed that patients perceived a benefit from having
increased salivary flow. The results were discussed with the FDA, which
suggested that the demonstrated improvement in saliva flow was encouraging, but not enough for marketing approval because saliva flow was a
secondary, not a primary, study endpoint. The FDA suggested that another
large-scale Phase III study be conducted with UWS flow as the primary
endpoint.

Because the cost of another study would be another six million dollars,
no further work was done.

***

A clinical trial in fibromyalgia was conducted in March 2000 and showed
relief of stiffness upon waking. Fibromyalgia syndrome had been recognized as a separate clinical entity, much to the relief of patients suffering
from chronic pain and stiffness who were told that fibromyalgia is a phony
diagnosis. The market in the United States has been estimated at five to ten
million patients, which may well be underestimated.

The diagnosis and recognition of fibromyalgia in Europe is not as
well established as in the United States, but it is thought that fibromyalgia in Europe is at least as prevalent as in America. Current treatment is
usually antidepressants and painkillers. There is a loss of working hours
in the United States due to fibromyalgia, and a heavy financial burden to
patients, employers, and the government (estimated at fifteen billion dollars
annually).

In our first fibromyalgia study, low-dose oral human interferon relieved patients from stiffness upon waking ("morning stiffness"), as published in the *Journal Interferon and Cytokine Research*:

> The study showed that the severity of morning stiffness was significantly reduced with the 50 IU/day when compared to the placebo group. An average 45% reduction in the average severity of the morning stiffness observed among non-Hispanic Caucasians was not trivial, especially since some subjects had much greater relief than the mean. In fact, the change in morning stiffness meets the 30% criteria for clinically relevant improvement.[144]

Patients reported feeling better than they had in years. Such results bode well for substantial market penetration in this important disease area.

The second clinical study also showed promising results. Patients were divided into three groups, and each patient took three lozenges per day. The three lozenges given to the first group contained 50 IU of HBL human interferon each, the second group was given one 50 IU HBL human interferon lozenge and two placebos, and members of the third group received three placebos. All groups reported a reduction in morning stiffness, but the improvement was most pronounced in those taking one 50 IU lozenge of HBL-human interferon daily.

However, the results did not reach statistical significance relative to the controls.

Participants were also given a low dose of the antidepressant drug amitriptyline, which they began taking one month prior to human interferon and continued throughout the three-month study of human interferon. The addition of the amitriptyline was necessary so the placebo patients would not have to tolerate a four-month period without any beneficial therapy. However, use of amitriptyline complicated the analysis and interpretation of the study results.

Patients who did not worsen during the first month's treatment with amitriptyline had a significant (p=0.0035) reduction in morning stiffness (when they took the 50 IU HBL human interferon lozenges once a day for three months, compared with placebo). However, those patients who reported worsening of their morning stiffness during the first month showed no benefit during three months of human interferon treatment.

We believed that a modified study design would confirm the therapeutic benefit of low-dose oral HBL human interferon in the treatment of morning stiffness in fibromyalgia patients.

But the lack of funding caused the studies to halt.

***

Deborah Greenspan, professor at UC San Francisco, treated oral complications of HIV-positive patients for thirty years, since the AIDS epidemic
began. Dr. Greenspan is considered the grande dame of HIV-positive
stomatology and the world's expert on oral warts. Although plantar warts
and warts on the hands and body come and go, warts in the mouth of
HIV-positive patients do not regress. Like herpesvirus, oral warts stay for a
lifetime in HIV-positive patients, we were told.

Dr. Greenspan has treated more than twenty HIV-positive patients with
human interferon lozenges. She noted regression of mouth warts—sometimes 100 percent regression in some of her patients given human interferon
lozenges. She has not found a treatment for her patients as safe and effective
as human interferon lozenges and thus became the strongest advocate and
proponent of human interferon lozenges for oral warts.

An open-label study was conducted in fifteen HIV-seropositive males
(aged thirty-five to fifty-seven years) with multiple oral warts. All men took
at least one protease inhibitor (PI) and two nucleoside or nonnucleoside
reverse transcriptase inhibitors for at least thirty days before enrollment in
this study.

Men were given HBL human interferon (150 IU lozenge) during weeks
zero through eight, and 450 IU/day (150 IU lozenges three times daily)
during weeks nine to sixteen. At week sixteen, the number of warts and
total surface area of warts were examined and compared to those measured
at baseline. In the men with an initial positive response, defined as more
than 10 percent reduction in total surface area of warts, human interferon
treatment was continued up to week forty. The protocol stated that treatment would be discontinued in men who did not show positive response at
week sixteen.

The investigator allowed any patient who wanted to remain in the study
to receive the drug up to week forty, since no other effective treatments
existed. One patient was lost to follow-up after the baseline visit. This is
what was reported:

> In this pilot study of fifteen HIV+ patients with multiple oral warts who
> were receiving HAART [highly active antiretroviral therapy], the use of low
> dose oral human interferon appeared to be safe and was associated with a

meaningful (>50%) reduction in the number and size of the oral warts in one-third of treated participants.[145]

In the other fourteen enrolled subjects, the mean total surface area (mm²) of warts was reduced from 1010.5±052.2 at baseline to 903.4±887.2 at week eight and 827.2±643.4 at week sixteen. One man with an initial response of a 23 percent decrease in wart area was lost to follow-up after week eight. At week sixteen, two men chose to stop treatment due to the lack of positive response.

Another man dropped from the study at week sixteen despite a 45 percent decrease in wart area.

Ten men continued HBL human interferon treatment beyond week sixteen, and eight men completed the forty-week treatment course. In these eight men, the total surface area (mm²) of warts was reduced from 1145.2±1377.9 at baseline to 282.8±381.2 at week forty (p=0.11). Total number of warts was significantly (p<0.03) decreased in these eight men from 20.4±11.1 at baseline to 9.0±6.9 at week forty. Two men stopped treatment at weeks thirty-two and thirty-six, respectively, because neither was a responder.

Among the fourteen evaluable men, one showed a complete response (100 percent clearance), four had a partial response (>50 percent clearance), two had a minor response (>25 percent clearance), and seven had no response. CD4+ counts and plasma viral loads tested in the study did not show any significant changes.

Based on the results of the pilot trial, a blinded, dose-ranging study was conducted by Atrix Laboratories, Inc. (Atrix) of Fort Collins, CO. Twenty-one HIV-seropositive men with multiple oral warts, who were receiving combination anti-retroviral therapy, were enrolled in groups of seven men to one of three treatment arms: 450 IU, 900 IU, or 1,500 IU human interferon per day.

Due to business reasons, the study was halted prematurely by Atrix, so not all men completed a treatment course. Nevertheless, a clear trend in favor of the HBL human interferon (1,500 IU) treatment arm was evident from this study. The overall response rates by group:

1.   450 IU: zero complete, one partial, two minor, and four nonresponders
2.   900 IU: one complete, one partial, zero minor, and five nonresponders

3. 1,500 IU: two complete, three partial, zero minor, and two
    nonresponders

Another study was a blinded, dose-ranging study in which twenty-one
men were enrolled in groups of seven men to one of three daily doses of
human interferon: 450, 900, or 1,500 IU/dose. After sixteen weeks of treatment, men were evaluated for initial response (≥10 percent reduction in wart
area). Men meeting this response criterion were eligible for twenty-four
additional weeks of treatment, but nonresponders were withdrawn.

Nearly all the adverse events reported in this trial were mild and
self-limiting.

Neither increased incidence nor severity of adverse events was noted at
higher doses, compared to HBL human interferon at 450 IU.

The most common adverse events noted were flu-like symptoms (five
men), oral Candidasis (two men), and diarrhea (two men).

Other studies reported that some oral warts regressed when HIV+
patients were given placebo. The "expert" opinion that oral warts do not
regress was questioned and made more clinical studies in this indication
less attractive. Our study in Los Angeles (Dr. Wilbert Jordan) revealed that
low-dose oral human interferon is safe and efficacious against anal warts
and anal cancers (both probably associated with papillomaviruses).

As previously mentioned, in October 1978, I took a single oral dose of
bovine interferon. Within three days, a wart on my left index finger, probably caused by papillomavirus, looked different. A red ring appeared around
the base of the wart, and the wart regressed in about three weeks. The wart
had been on my finger for at least ten years.

Over the past decades, others using oral human interferon reported that
warts regressed. Skin warts are not the same thing as cancer caused by papillomavirus, but the same underlying immune response may be effective
against both.

After fifty years of human interferon research, this was another of many
examples where low-dose application worked as well as, or better than, high-
dose injected human interferon.

***

Behçet's disease (BD) is a severe, rare, chronic, relapsing autoimmune disorder characterized by oral and genital ulcers, eye inflammation (uveitis),
and skin lesions, as well as joint, blood vessel, central nervous system, and

gastrointestinal tract inflammation. The oral lesions occur in all patients at some time in the disease. BD is found worldwide and is a significant cause of disability. The US patient population has been estimated at fifteen thousand. The FDA Office of Orphan Product Development granted orphan drug status for low-dose orally administered human interferon in this condition. "Orphan drug" status is granted to those treatments that appear to be safe but don't have enough patients for the pharmaceutical companies to justify the money needed for clinical trials.

At the end of February 2006, Martin Cummins, vice president of Regulatory and Clinical Affairs, visited Nobel Ilac Sanayii Ve Ticaret A.S., our licensee in Turkey. He participated in study initiation meetings with the clinical investigators prior to the commencement of enrollment of eighty-four patients with BD in a study of human interferon lozenges versus placebo.

Anecdotal evidence indicated that low-dose oral administration of human interferon was beneficial in the management of recurrent oral ulcers (OU). A double-blind, placebo-controlled study was initiated to explore the efficacy and safety of low doses of natural human interferon in the treatment of OU in BD patients.

A total of seventy-four (fifty males, twenty-four females) patients with a history of recurrent OU were enrolled at four treatment centers in Turkey. Patients exhibiting at least two OU were allocated to daily treatment with human interferon at either 1,000 IU or 2,000 IU, or placebo. Subjects were monitored weekly for the first four weeks and every other week for the next eight weeks of treatment. OU were counted and measured at the study visits. The primary efficacy end point was the difference in the total OU at week twelve compared to that at week zero.

Of seventy-four patients, seventy-two completed the trial. The dropout rate was similar among the groups and there were no statistically significant differences among the groups with respect to the number of patients reporting an adverse event or with respect to any efficacy endpoints.

While safety was confirmed, treatment with human interferon failed to reduce the total OU burden in these BD patients. Data generated over years indicate that either 1,000 IU or 2,000 IU of human interferon is too high a dose. The dose that will help treat BD is probably 50 IU, not these higher doses.

After the study in Turkey, we learned how critical it is to keep the dose of oral human interferon low to help down-regulate expression of autoimmune and inflammation genes.

Fifteen BD patients in the United States given low-dose natural bovine interferon reported clinical benefit (placebo was not tested). A woman in Tennessee has used low-dose natural bovine interferon for twenty years to manage BD (her story appears in the Testimonial section).

***

Idiopathic Pulmonary Fibrosis (IPF) is a chronic inflammatory fibrotic disease of the lower respiratory tract distinguished by alveolitis dominated by alveolar macrophages, polymorphonuclear leukocytes, lymphocytes, and eosinophils.

Regarding the use of the word "idiopathic," in plain language it simply means "We don't know what's causing the problem." It could be a virus, a bacterium, some other type of pathogen, or a genetic problem.

The disease presents as dyspnea on exertion, the chest x-ray shows diffuse reticulonodular infiltrates, and analysis of lung function reveals restrictive abnormalities. IPF is usually fatal in three to five years unless the IPF patient has a lung transplant. Bronchiolitis of respiratory bronchioles may be present, and alveolar units are always involved.

Low-dose orally administered HBL human interferon was tested as a treatment of IPF under an Advanced Technology Program Grant awarded by the State of Texas. The hundred thousand dollar grant was used by the Health Science Center to conduct a study of twenty IPF patients. ABI cooperated on this research with Lorenz Lutherer, MD, PhD, a professor of physiology; and Cynthia Jumper, MD, associate professor of patient care, in internal medicine. We provided support in the form of study drug, data management, and biostatistical analysis.

A trial of low-dose orally administered human interferon (150 IU three times daily) showed minimal adverse events. Patients were evaluated with pulmonary function tests every three months and high-resolution computed tomography (HRCT) at yearly intervals. Twelve patients completed treatment for at least one year; the forced vital capacity remained stable in eight patients, and the oxygen saturation after a six-minute walk was stable in seven and improved in one patient. This is what was reported:

> Clinical data on the 12 subjects who completed at least 1 year of treatment are summarized in table 1. All subjects tolerated treatment well. Using the criteria from the International Consensus Statement, FVC was stable in 10 subjects (12 evaluable), and O2 saturation post-exercise was stable or improved in nine

subjects (11 evaluable) over a 12-month period. High resolution CTs (HRCTs) showed no evidence of progression after 1 year in seven subjects (11 evaluable) and only slight progression in the other four. Two subjects followed for 36 and 57 months showed stability on the PFTs [pulmonary function tests] and no progression on the HRCT.[146]

One patient showing lack of progression was followed for over five years, and another subject was followed for three years. The eight patients whose pulmonary function tests were stable showed no evidence of disease progression on HRCT scans.

Most patients who entered the study with a cough noted marked improvement within the first few weeks of treatment with corresponding increases in quality of life scores. These results strongly suggest that low dose oral human interferon can prevent progression according to the criteria defined by the American Thoracic Society.

A woman in Texas, diagnosed with IPF thirty years ago, died January 2018 after years of successfully using human interferon. A man diagnosed with IPF for twelve years died. Both people attributed their remarkable long-term survival to low-dose oral human interferon. The cost and safety of low-dose oral human interferon make it a reasonable treatment for IPF.

***

Low doses of oral human interferon are safe and effective in many species and disease conditions.[147]

Horses with inflammatory airway disease weighing a thousand pounds responded (P<0.05) to a daily oral dose of HBL human interferon at 50 IU, but not at 450 IU.

Cattle with respiratory tract disease experience significant (P<0.01) improved survival if given a single oral dose of human interferon (less than 1.0 IU per pound of body weight), compared to placebo.[148]

Pigs, dogs, cats, mice, rats, and poultry can be given low doses of human interferon orally or intranasally, and beneficial systemic effects are observed. Low doses of human interferon probably do not have a direct antiviral effect but instead exert an immune modulatory effect through human interferon-stimulated genes.[149]

Clinical testing of low-dose oral human interferon in blinded con-trolled studies of 370 human volunteers challenged with rhinovirus failed to demonstrate a benefit.[150] Indeed, there are published reports in which oral or intranasal human interferon was not useful in animals.

But the observations from thousands of influenza patients should not be dismissed. Rhinovirus does not share the same viral classification with influenza virus, and the pathobiology and response to human interferon are quite different.

In view of the serious consequences of an influenza pandemic and because of our present understanding of the ability of small doses of human interferon to modulate immune functions, it is time to revisit the subject of low-dose oral or intranasal human interferon for influenza therapy and/ or prophylaxis. Perhaps the Soviets and others were correct in their clinical observations.

It is time to put these old observations to the test, not by using high doses of oral or intranasal human interferon, but by using low doses of human interferon of high purity.

Studies of oral human interferon have been conducted around the world and should not be ignored. I believe science has an important tool to fight infectious diseases, and it is being overlooked.

We have tried to tell scientists, WHO, NIH, CDC, newspapers, politicians, doctors, and others that low-dose oral or intranasal interferon is safe and effective in many conditions. In the next chapter, we will talk about interferon and what help it might provide against COVID-19 and other coronaviruses.

CHAPTER 10

Interferon and COVID-19?

I want you to recall the beginning of this book and the *TIME* magazine cover story from March 31, 1980. One of the sections highlighted was how interferon showed promise in treating the common cold. The common cold is a coronavirus, like SARS-CoV-2, which causes COVID-19. (Rhinoviruses also often cause what we call "the common cold.") This is directly from the CDC website:

> Common human coronaviruses, including types 229E, NL63, OC43, and HKU1, usually cause mild to moderate upper-respiratory tract illnesses, like the common cold. Most people get infected with one or more of these viruses at some point in their lives.[151]

You're also probably aware that even though the common cold is, well, common, we haven't developed a vaccine against it. That's because it's extremely difficult to develop a vaccine against a coronavirus. Even though President Trump wants to develop a vaccine against SARS-CoV-2 at "warp speed," I'm very skeptical that he will succeed. In fact, as an alternative, I think low-dose interferon may be a substitute for any vaccine. Maybe those well-baby visits, instead of injections, might just have a dropperful of interferon given to the child, and nobody has to worry about heavy metals, animal viruses from the manufacturing process, or negative side effects.

A publication from 1987 in the journal *Antiviral Research* looked specifically at the question of interferon, coronavirus, and activation of a hepatitis virus in a mouse model of disease. The authors began by observing:

Intranasally administered recombinant alpha interferon (IFN) has been shown to reduce upper respiratory clinical signs, nasal secretions and virus excretion after challenge of human volunteers with coronavirus 229E (Higgins et al., 1983). In that study, intranasal recombinant IFN treatment shortened the duration and reduced the severity of cold symptoms following challenge with coronavirus 229E (Turner et al., 1986).[152]

As a veterinarian, I must admit my bias that humans are just big animals. I understand each species will have unique characteristics, but general principles apply. Notice here how the research first conducted on humans was later applied to animals. The findings among the mice with coronavirus were quite remarkable, as detailed in the discussion section. In addition, this research clearly demonstrated the usefulness of lower doses of interferon:

Based on the data presented here, we conclude that interferon doses approximately 10-fold lower per unit weight than those reported in human clinical trials (Higgins et al., 1983; Turner et al., 1986) alter the course of murine coronavirus disease. Intranasally administered interferon had a substantial effect on local virus replication in the nose and modified the process of direct virus extension from olfactory tissue to the brain.[153]

In 1994, I published an article in collaboration with several other researchers on the use of natural human interferon among pigs who were suffering from an outbreak of transmissible gastroenteritis (TGE) virus, a coronavirus:

The TGE epidemic described in this study is typical of the disease in a high-production closed farrowing facility. None of the sows were vaccinated prior to the outbreak, and thus, pigs of all ages were immunologically naïve and susceptible to TGE virus. Although the exact start of the TGE outbreak could not be determined, the epidemic was well-established by the time therapy was initiated. The survival data were collected and analyzed by three age groups at initiation of human interferon therapy. The most beneficial effect of human interferon therapy were observed in 1-12 day old piglets; all three dosages of human interferon were effective. In contrast, newborns that were farrowed during the disease outbreak, and treated with human interferon within a few hours of birth, did not demonstrate a significant treatment response.[154]

It's important to understand the difficulties under which we were operating in this study. This outbreak among the pigs had already started when I came up with the idea to use interferon. Let me share how I felt when I got on-site and was able to assess the situation.

I had never seen TGE and cannot recall why I wanted to do such a study. Some friends identified a natural TGE outbreak in swine in Pennsylvania, so I flew there from Texas with human interferon. Upon arrival at the pig farm, I could not believe my eyes and nose. What a disaster. The piglets were vomiting and had diarrhea.

I got on my knees and started treatment of 1,740 piglets orally for four consecutive days. Some pigs were only treated one, two, or three days before they died. I had placebo and three different doses of human interferon but was unable to split piglets in the 203 litters, so a litter stayed on the same human interferon dose for four days. Piglets were born into this epidemic and were treated as newborns, and other piglets one to twenty days old were treated.

I recall thinking in those first days I was sorry I got involved in such a mess. I was sure that under these terrible conditions, I was wasting everyone's time and could not impact the death loss. Imagine my delight when the data were analyzed and a significant survival benefit was noted.

The piglets in the one-to-twelve-days age group increased their chances of survival from 15.2 percent to 50 percent with the use of interferon, an approximately 3.2-fold increase.

The piglets that were from thirteen to twenty days old were able to survive (>90%) whether given interferon or placebo.

The piglets given interferon within a few hours of birth did not increase their survival chances.

I believe this is strong evidence that the immune system needs to be somewhat developed in order for interferon to be most effective.

***

In the current crisis we face with the SARS-CoV-2 virus causing COVID-19, a critical question is whether the immune system will recognize the virus and produce antibodies to help fight the disease. A recent publication in *Science* magazine suggests that those who have previously been exposed to SARS-CoV-2 will have strong protection from any further exposures, and even those who have not will also have some measure of protection, perhaps from exposure to other viruses, such as the virus that causes the common cold:

> Immune warriors known as T cells help us fight some viruses, but their importance for battling SARS-CoV-2, the virus that causes COVID-19, has been unclear. Now, two studies reveal infected people harbor T cells that target the virus—and may help them recover. Both studies also found some people never infected with SARS-CoV-2 have these cellular defenses, most likely because they were previously infected with other coronaviruses.[155]

Just for a quick lesson, the T cells are probably the most important cells of the immune system and function in two important ways. They will configure themselves in a manner to directly attack viruses (called "killer T cells") and other pathogens, as well as other T cells (called "helper T cells") directing the B cells of the immune system into action. The article continued:

> Using bioinformatics tools, a team led by Shane Crotty and Alessandro Sette, immunologists at the La Jolla Institute for Immunology, predicted which viral protein pieces would provoke the most powerful T cell responses. They then exposed immune cells from 10 patients who had recovered from mild cases of COVID-19 to these viral snippets.
>
> All of the patients carried helper T cells that recognized the SARS-CoV-2 spike protein, which enables the virus to infiltrate our cells. They also harbored helper T cells that react to other SARS-CoV-2 proteins. And the team detected virus-specific killer T cells in 70% of the subjects, they report today in *Cell*. "The immune system sees this virus and mounts an effective immune response," Sette says.
>
> The results jibe with those of a study posted as a preprint on medRxiv on 22 April by immunologist Andreas Thiel of the Charité University Hospital in Berlin and colleagues. They identified helper T cells targeting the spike protein in 15 out of 18 patients hospitalized with COVID-19.[156]

One should consider the helper T cells to be the alarm system of the body, correctly identifying threats. In the bodies of those who'd previously had a SARS-CoV-2 infection, 100 percent of these cells correctly identified the virus. In addition, 70 percent of the entire group also had specific killer T cells already created to deal with the threat. The research from the La Jolla institute was also consistent with an earlier study from Germany. The last section from this article highlighted findings regarding how the immune system of individuals never exposed to SARS-CoV-2 respond:

> The teams also asked whether people who haven't been infected with SARS-CoV-2 also produce cells that combat it. Thiel and colleagues analyzed blood from 68 uninfected people and found that 34% hosted helper T cells that recognized SARS-CoV-2. The La Jolla team detected this cross-reactivity in about half of stored blood samples collected between 2015 and 2018, well before the current pandemic began. The researchers think these cells were likely triggered by past infection with one of the four human coronaviruses that cause colds; proteins in these viruses resemble those of SARS-CoV-2.[157]

The German team found that 34 percent of the helper T cells in those who'd never been exposed to SARS-CoV-2 correctly recognized it. The La Jolla team found that number to be close to 50 percent. There seemed to be a fairly strong cross-reactivity from the normal coronaviruses to which we might be exposed, such as the common cold.

Now, might interferon be able to make our immune system work even better to respond to SARS-Cov-2?

***

Research from the University of Texas in the *Journal of Antiviral Resistance* in April 2020 showed that interferon was highly effective against SARS-CoV-2 in human cell cultures. The researchers reported:

> Our data clearly demonstrated that SARS-CoV-2 is highly sensitive to both IFN-*a* and IFN-*b* in cultured cells, which is comparable to the IFN sensitive VSV. Our discovery reveals a weakness of the new coronavirus, which may be informative to antiviral development . . . Our data may provide an explanation, at least in part, to the observation that approximately 80% of patients actually develop mild symptoms and recover. It is possible that many of them are able to mount IFN-*a/b*-mediated innate immune response upon SARS-CoV-2 infection, which helps to limit virus infection/dissemination at an early stage of the disease.[158]

Let's consider the size of the problem posed by SARS-CoV-2. Eighty percent of those who become infected will develop only mild symptoms and will develop the necessary antibodies that will likely protect them in case of any further exposure.

When the scientists added both interferon-alpha and interferon-beta to the cultured cells, this is what they found:

> We next tested the antiviral efficacy of IFN-*a* and IFN-*b* at lower concentrations. (1-50 IU/ml). Both IFN-*a* and IFN-*b* dose-dependently inhibited virus infection at these lower concentrations. IFN-*a* exhibited anti-SARS-CoV-2 activity at a concentration as low as 5 IU/ml, resulting in a significant reduction of viral titer by over 1 log ($P<0.01$). With increasing IFN-*a* concentrations, the viral titers steadily decreased. Treatment with IFN-*a* at 50 IU/ml drastically reduces viral titers by 3.4 log. Treatment with 1 IU/ml of IFN-*b* resulted in a moderate (approximately 70%) but significant decrease in virus titer ($P,0.05$, student *t* test). Infectious virus was nearly undetectable upon treatment with 10, 25, and 50 IU/ml of IFN-*b*.[159]

There's probably a good explanation in this example why interferon-*b* seems to have a slightly greater effect than interferon-*a*. But who really cares what the difference is if we can stop the disease? Just so one understands the use of the word "log" in this concept, it stands for the power of ten. Thus, a one-log reduction means the amount of the virus has been reduced ten-fold. A 3.4-log reduction means the virus has been reduced thirty-four-fold. Reducing the viral load is one of the best strategies to cut the danger of a virus. In plain English, there's just not that much virus there. The body can deal with it.

The article concluded this section by stating, "Taken together, these results indicate that treatment with low concentrations of both IFN-*a* and IFN-*b* significantly inhibited viral infection, with IFN-*b* being slightly more effective than IFN-*a*."[160]

Researchers from Hong Kong published in *The Lancet* on May 30, 2020, their findings of combining low-dose interferon (50 IU/ml) with two other antiviral drugs, ribavirin and lopinavir-ritonavir in human subjects. The study consisted of 126 patients, and this is what they found:

> In this multicenter randomized open-label phase 2 trial in patients with COVID-19 in patients with COVID-19, we showed that a triple combination of an injectable interferon (interferon beta-1b), oral protease inhibitor (lopinavir-ritonavir), and an oral nucleoside analogue (ribavirin), when given within 7 days of symptom onset, is effective in suppressing the shedding of

SARS-CoV-2, not just in nasopharyngeal swab, but in **all** (bold and under-line added) clinical specimens, compared with lopinavir-ritonavir alone.

Furthermore, the significant reductions in duration of RT-PCR positiv-ity and viral load were associated with clinical improvement as shown by the significant reduction in NEWS2 and duration of hospital stay. Most patients treated with the triple combination were RT-PCR negative in all specimens by day 8. The side effects were generally mild and self-limiting.[161]

These were extraordinary results. The triple combination, if given within seven days of the onset of symptoms, resulted in a complete repres-sion of ALL viral shedding within eight days (meaning there is no possible way the patient could spread the virus) and that most were negative in their tests for the virus.

In addition, it seems that interferon was the key factor in the positive results. They did not get the same results with just lopinavir-ritonavir and ribavirin. And the side effects were "generally mild and self-limiting." Now, there were some differences, such as the interferon was injected, and while I would prefer it to be given by mouth or nasal spray, this method would still be effective.

In Iran, researchers looked at the question of interferon and COVID-19, without any other medications. There were some problems with the study, such as their using high levels of interferon (12 million IU/ml) that were injected, but the results were encouraging. These were their results:

As primary outcome, time to the clinical response was not significantly dif-ferent between the interferon (IFN) and the control groups (9.7 +/- 5.8 vs. 8.3 +/- 4.9 days respectively, P=0.95). On day 14, 66.7% vs.43.6% of patients in the IFN group and the control group were discharged, respectively (OR = 2.5; CI: 1.05 – 6.37). The 28-day overall mortality was significantly lower in the IFN then the control group (19% vs. 43.6% respectively, p = 0.015). Early adminis-tration significantly reduced mortality (OR+13.5; 95% CI: 1.5-118).[162]

Even though the Iranian researchers gave far too much interferon and injected it, they were still able to cut the mortality rate by a little more than half (43.6 percent to 19 percent). I believe that if they'd given low doses of interferon, the results would have been much more in line with the Hong Kong researchers, in which a complete repression of viral shedding after eight days was reported. The Hong Kong researchers gave 50 IU/ml of inter-feron, while the Iranian researchers gave 12 million IU/ml of interferon.

Low-dose interferon is likely to be effective against SARS-CoV-2 and COVID-19, or any other viral disease we might encounter in the future.

***

Another study that should probably be discussed is one from a different group of Hong Kong researchers who used interferon without any other medications. Prior to publication in the journal *Frontiers in Immunology* on May 15, 2020, the findings were evaluated and approved by researchers from Walter Reed Army Institute of Research, the National Institutes of Health, and Johns Hopkins University.

There were some shortcomings to the research, conducted as it was during a pandemic, but the results were extremely promising in the seventy-seven patients treated. The authors found:

> This uncontrolled, exploratory study provides several important and novel insights into COVID-19 disease. Importantly, human interferon alpha 2 b (IFN-*a*2b) therapy appears to shorten duration of viral shedding. Reduction of markers of acute inflammation such as CRP and IL-6 correlated with the shortened viral shedding, suggesting IFN-*a*2b acted along a functional cause-effect chain where virally induced inflammation represents a patho-physiological driver. Taken together, these findings support the [use] of IFN-*a*2b [as] a therapy for COVID-19 disease.[163]

It should be noted that in COVID-19, as in many diseases, much of the damage seems to come from the body's inflammatory response. If you control the inflammation, you can control the damage, especially to the organs. The authors went on to explain:

> The reduction of the inflammatory biomarker IL-6 following inhaled IFN-*a*2b therapy not only supported a clinically relevant impact of the approach, but also hinted at likely functional connections between viral infection and host end organ damage. IL-6 has been shown to provide prognostic value in acute respiratory disease syndrome (ARDS), which is the most severe form of COVID-19 disease.[164]

IL-6 is an important inflammatory marker and serves as an indicator that the danger is beginning to lessen. Low IL-6 indicators means the possibility of organ damage goes down dramatically. The authors were cautious in their findings, suggesting that larger, better-designed studies needed to be undertaken, but still noting:

> Irrespective of thee significant limitations, to our knowledge, the findings presented here are the first to suggest therapeutic efficacy of IFN-a2b, an available antiviral intervention. Furthermore, beyond clinical benefit to the individual patient, treatment with IFN-a2b may also benefit public health measures aimed at slowing the tide of this pandemic, in that duration of viral shedding appears shortened.[165]

Science is supposed to be about improving the health of humanity. It is not about ego. I may have played some small part in the development of interferon, but my time on the stage is just about finished. I'm an old man, I have Parkinson's disease, and I'm not interested in money. I'm just interested in making a contribution. I did not in any way collaborate or advise on this study, but I'm not surprised by the results.

Interferon can help the individual patient, and it may provide broad protection to the public by shortening the time of viral shedding. If we are to endure future outbreaks of viral diseases, perhaps interferon should be our first line of defense.

***

It's been said that human beings always do the right thing, but only after doing all the wrong things first.

My opinion is we've been making mistakes with interferon for more than forty years. I've often been disillusioned, wondering if there were things I should have done differently. But maybe all of this should have been expected, especially when interferon threatens to upend the pharmaceutical bottom line. In science there's a joke that knowledge doesn't advance by convincing the current professionals of a new way of thinking with new data. Instead, goes the joke, "science advances one funeral at a time," requiring the old guard with their old ideas to die off, and those with a different viewpoint to obtain power in their academic realms.

As I mentioned in the Introduction, the general public was first introduced to the concept of interferon in a 1960 Flash Gordon comic strip in which it was used to fight off a newly discovered space virus. Like many Americans, I was thrilled to watch our country return to space on May 30, 2020, with the launch of Elon Musk's Falcon 9 Dragon capsule carrying a crew to the International Space Station. Musk has declared his intention to build the first city on Mars and make us an interplanetary species.

When we go to Mars to establish our first colony on another world, we will be confronting unimaginable dangers. Space is not forgiving, and the Martian terrain will not be much better. We will need to be strategic in what we bring, as we will eventually need to figure out how to live off the land. Perhaps that means we will build great domes under which to live or excavate great underground caverns to spare us from the merciless solar radiation on the Martian surface.

We will also need to be strategic in what we bring to Mars to confront whatever health problems we might encounter in this new world. The brave astronauts who settle Mars are likely to be the most physically fit among us, so we need only worry about keeping their immune system strong.

It is my hope that when man leaves Earth to settle this new frontier, interferon will be in the medicine cabinet of the first city on Mars.

Testimonials

Behçet's Disease
by Dorothy "Dot" Tutt, Knoxville, TN

I have an immune disorder called Behçet's Disease. I have a hyper-reactive immune system and all of my body systems are involved. I am in my late sixties now but spent my life experiencing wild, weird, unexplainable symptoms from the time I was a toddler until my late forties, when I was finally diagnosed, and the search for a safe treatment began. I suffered adverse reactions to not only umpteen foods, fumes, and chemicals, but I also reacted to almost all medications/treatments. My father, diagnosed posthumously with the same disease, had already died from side effects of the methotrexate the doctors had prescribed to try to suppress his immune reactions. We couldn't find a safe and effective treatment for me using standard medications.

I was at death's door when I tried Dr. Cummins's low dose oral natural human interferon lozenges. Despite having passed clinical trials in the United States with flying colors, it did not receive approval by the FDA due to the covert objections of pharmaceutical companies that were working to protect their own profits derived from sales of injectable interferon. Dr. Cummins's product was just as effective but much less expensive and had been proven to not cause the adverse reactions caused by the injectable version. It was approved in Canada, so I had to apply to the FDA for a personal use exemption to import it from Canada.

At the time, I was unable to walk due to muscle and joint involvement; I could not talk (aphasia) or think ("brain fog") due to lesions in my brain; I had constant severe headaches; I was bleeding through my retinas; I had deep ulcers throughout my digestive system from my mouth through my intestines causing constant pain and severe diarrhea; my hair was brittle and falling out; I had skin lesions; and many more symptoms. All other doctors had given up on me; but my brilliant internist was able to diagnose me and started my treatment with Dr. Cummins's low dose oral natural human interferon lozenges. I experienced major benefits within two days, most of my symptoms were gone within a week, and a few lesser symptoms (tinnitus and painful breast lumps) took almost three months to go away. I still have some lingering issues like permanent joint damage, deafness in my left ear, scar tissue in my lungs and intestines, etc., which were caused by my disease before I was successfully treated.

The company producing the lozenges went out of business so I switched to Dr. Cummins's low dose oral natural bovine interferon liquid ("bovine") that was very effective in tiny doses. By using that product, over time I was able to desensitize my immune system to most things to which I used to react.

Both of our sons were also diagnosed with Behçet's and all four grand-children also exhibit signs of Behçet's. Our sons and I have all reached a state of relative remission using Dr. Cummins's natural interferon products, so we no longer take it daily, but we still use bovine occasionally when we are accidentally exposed to something that would have normally set off a major allergic reaction or dangerous vasculitis attack in the past. Our grandchildren similarly use it to control their symptoms as needed.

Since we have personally found these products also do an amazing job helping us fight off colds and flu, we plan to use bovine to battle the coronavirus should we become infected. Our sons think they have both already successfully fought off COVID-19 using it, but it took so long to get approved for a coronavirus test that they were already better before they could get tested.

I highly recommend the low dose oral natural bovine interferon. Without it, both of my sons and I would be dead and possibly our grand-children. I personally cannot survive without it.

Fever Blister Story
by New Mexico resident Crystal Sebring

I have always gotten fever blisters since I was an infant. They come and go. I have gone a couple years before without getting a fever blister, but I have never had such a consistent and recurring outbreak like I have had the last three years. It started around the fall of 2013—maybe it was stress, being in graduate school, hormone changes, or perhaps a combination of things—but for whatever reason, I started getting fever blisters with a vengeance.

Almost like clockwork, I would get one to three fever blisters once a month, approximately around my monthly cycle . . . each and every month for the last three years. By the time one would heal, I would have a fever blister-free few days before the next one popped up. I tried every home remedy including toothpaste, corn starch, and coconut oil. I tried medications. I took a round of prescription pills for the herpes simplex virus, and I tried over-the-counter Abreva, Lysine, etc.

Last summer I started getting them in my nose, as well. Fed up, I went back to the doctor, who prescribed me a cream . . . an eight-hundred-dollar cream that my insurance denied. So after getting yet another blister in August 2016, I decided to try natural bovine interferon diligently. I take three to four sprays in the morning and three to four sprays at night.

I am happy to say that I have now gone seven months without a fever blister! This may not seem like a long time, but when you get fever blisters every month like clockwork, it's very exciting! I have also noticed that my illnesses have a shorter duration and less intensity than before. I work with children, so I do get sick, but I also have not had to take any antibiotics so far this school year or had severe bronchitis like I usually get. My "colds" last a few days instead of a few weeks.

I will continue to take natural bovine interferon daily and keep track of my days without a fever blister.

COPD Story
of Earl S.

Earl S. is a man (eighty-seven years old) living in Oklahoma. Earl began smoking cigarettes when he was fifteen years old. When Earl was smoking four packs of cigarettes a day about forty years ago, he stopped smoking.

About twelve years ago, Earl was diagnosed with Chronic Obstructive Pulmonary Disease (COPD) by a pulmonologist at an OKC Oklahoma City lung institute. Recently, the same OKC lung institute conducted extensive lung function tests and concluded that Earl did not have COPD. Earl was told by the attending pulmonologist that Earl "had the lungs of a sixty-year-old man who had never smoked."

Earl is an old friend of mine who has had access to oral human and bovine interferon for at least two decades. For the past six years, he has used bovine

interferon exclusively. For the past four years, Earl has taken this bovine interferon faithfully twice daily.

***

As I edit this history in October 2020, I note that our research on human
interferon and bovine interferon has reached its fiftieth anniversary. The
low-dose oral human interferon technology is easy to administer, safe, and
inexpensive. If our technology is ever accepted, it will make those who care
for animals and humans look good.

How can it fail to gain acceptance?

Endnotes

Introduction

[1] "INTERFERON: The Big IF in Cancer," *Time*, March 31, 1980, pp. 60–64.

[2] Ibid., 64.

[3] Ibid.

[4] Ibid.

[5] Ibid.

[6] Michael Edelhart, "Putting Interferon to the Test," *New York Times* magazine, April 26, 1981.

[7] Ibid.

[8] Ibid.

[9] Ibid.

[10] Ibid.

[11] Ibid.

Chapter 3

[12] V.D. Soloviev, "The Results of Controlled Observations on the Prophylaxis of Influenza with Interferon," *Bulletin of the World Health Organization*, Vol. 41, pp. 683–688 (1969).

[13] Ibid., 684.

[14] Ibid., 688.

[15] V.D. Soloviev, *The Interferons*, Academic Press, New York, pp. 233–243 (1967).

[16] "Antiviral Research in the Soviet Union," *The Journal of Infectious Diseases*," Vol. 125, Issue 4, pp. 455–456 (April 1972).

[17] William S. Jordan, Jr., Hope E. Hopps & Thomas C. Merigan, "Influenza and Interferon Research in the Soviet Union: January 1973," *The Journal of Infectious Diseases*, Vol. 128, No. 2, pp. 261–264 (August 1973).

[18] Ibid.

[19] Thomas C. Merigan, Thomas S. Hall, Sylvia E. Reed & David A.J. Tyrell, "Inhibition of Respiratory Virus Infection by Locally Applied Interferon," *The*

Lancet, Vol. 301, Issue 7803, pp. 563–567 (March 1973): doi: 10.1016/S0140–6736 (73)90714–9.

[20] Veselina Arnaoudova, "Treatment and Prevention of Acute Respiratory Virus Infections in Children with Leukocytic Interferon," *Rev. Roum. Med.*, Vol. 27, Issue 2, pp. 83–88 (1976).

[21] June K. Dunnick & George J. Galasso, "Clinical Trials with Exogenous Interferon: Summary of a Meeting," *The Journal of Infectious Diseases*, Vol. 139, Issue 1, pp. 109–123 (January 1979): doi: 10.1093/infdis/139.1.109.

[22] J. Imanishi, T. Karaki, O. Sasaki, et al., "The preventive effect of human interferon-alpha preparation on upper respiratory disease," *J Interferon Res.*, 1980 Fall; 1(1):169–178.

[23] S. Isomura, T. Ichikawa, M. Miyazu, et al., "The Preventive Effects of Human Interferon-Alpha on Influenza Infection; Modification of Clinical Manifestations of Influenza in Children in a Closed Community," *Biken Journal*, Vol. 3, pp. 131–137 (September 25, 1982).

[24] Dai JiaXiong, You Chun-Hua, Qi Zhong-Tian, et al., "Children's Respiratory Viral Diseases Treated with Interferon Aerosol," *Chinese Medical Journal*, Vol. 100(2), pp. 162–166 (1987).

[25] Hitoshi Sato, Hiroshi Takenaka, Saicho Yoshida, et al., "Prevention from Naturally Acquired Viral Respiratory Infection by Interferon Nasal Spray," *Rhinology*, Vol. 23, pp. 291–295 (1985).

[26] R.J. Phillpotts, P.G. Higgins, J.S. Willman, et al., "Intranasal Lymphoblastoid Interferon ('Wellferon') Prophylaxis against Rhinovirus and Influenza Virus in Volunteers," *Journal of Interferon Research*, Vol. 4, pp. 535–541 (1984).

[27] J.J. Treanor, R.F. Betts, S.M. Erb, et al., "Intranasally Administered Interferon as Prophylaxis against Experimentally Induced Influenza A Virus Infection in Humans," *Journal of Infectious Diseases*, August 1987: 156(2): 379–83: doi: 10.1093 /infdis/156.2.379.

[28] Ibid.

[29] F.G. Hayden, A.N. Schlepushkin & N.L. Pushkarskaya, "Combined Interferon-Alpha 2, Rimantadine Hydrochloride, and Ribavirin Inhibition of Influenza Virus Replication in Vitro," *Antimicrobial Agents and Chemotherapy*, January 1984, 25(1): 53–57, doi: 10.1128/aac.25.1.53.

[30] G.M. Scott, R.J. Phillpotts, J. Wallace, et al., "Purified Interferon as Protection Against Rhinovirus Infection," *British Medical Journal*, June 1982, 19;284 (6332:1822–5: doi:10.1136/bmj.284.6332.1822.

[31] P.E. Came & W.A. Carter, "Interferons and their Applications," *Handbook of Experimental Pharmacology*, Vol. 71, p. 433 (1984), Springer-Verlag, New York, Tokyo, Berlin, and Heidleberg: doi:10.1002/jobm.3620250408.

[32] Kari Cantell, *The Story of Interferon: The Ups and Downs in the Life of a Scientist*, World Scientific Publishing, Helsinki, Finland, May 1998, doi.org/10.1142/3486.

[33] Ibid., 220.

[34] J. Imanshi, T. Karaki, O. Sasaki, et al., "The Preventive Effect of Human Interferon-Alpha Preparation on Upper respiratory Disease," *Journal of Interferon Research* (1980), Vol. 1, 169–178.

[35] S. Isomura, T. Ichikawa, M. Miyazu, et al., "The Preventive Effect of Human interferon-alpha on Influenza Infection: Modification of Clinical Manifestations of Influenza in Children in a Closed Community," *Biken Journal* (1982), Vol. 25: 131–137.

[36] H. Saito, H. Takenaka, S. Yoshida, et al., "Prevention from Naturally Acquired Viral Respiratory Infection by Interferon Nasal Spray," *Journal of Rhinology* (December 1985), Vol. 23(4): 291–5.

[37] Alayne Bennett, David Smith, Martin Cummins, et al., "Low-Dose Oral Interferon Alpha as Prophylaxis against Respiratory Illness: A Double-Blind, Parallel Controlled Drug Trial During an Influenza Pandemic Year," *Influenza and Other Respiratory Viruses* (February 9, 2013), Vol. 7, Issue 5, pp. 854–862; doi.org/10.1111/irv.12094.

Chapter 4

[38] Thomas Schaeffer, Melvn Lieberman, Mildred Cohen & Paul Crane, "Interferon Administered Orally: Protection of Neonatal Mice from Lethal Virus Challenge," *Science*, Vol. 176, Issue 4041, pp. 1326–1327: doi: 10.1126/science.176.4041.1326.

[39] L. Montevecchi, G. Caprio & A. Vecchione, "Preliminary Note on the Use of Interferon Alpha by Peroral Route in HPV Lesions," *Clin. Ter.* 151 (suppl. 1): 29–34 (2000).

[40] M. Palomba & G.B. Mellis, "Oral Use of Interferon Therapy in Cervical Human Papillomavirus Infection," *Clin. Ter.* (suppl. 1): 59–61 (2000).

[41] C. Bastianellie, M.T. Caruso & G.F. Marcellini, "New Treatment of Viral Genital Lesions with low Dosage of Interferon Alpha by Oropharyngeal Absorption," *Clin. Ter.* 151 (suppl. 1): 23–28 (2000).

[42] A. Biamonti, M. Cangialosi, R. Brozzo, et al., "Peroral Alpha-Interferon Therapy in HPV-lesions of the Lower Female Genital Tract: Preliminary Results," *Clin. Ter.* 151 (suppl. 1): 53–58 (2000).

[43] S. Verardi, E. Zupy, D. Marconi, et al., "Cutaneous and Mucous Infections from Human Papilloma Virus: New Therapeutic Approach," *Clin. Ter.* 151 (suppl. 1): 35–52 (2000).

[44] M.B. Tompkins & J.M. Cummins, "Response of Feline Leukemia Virus-Induced Nonregenerative Anemia to Oral Administration of an Interferon-containing Preparation," *Feline Practice*, Vol. 12, No. 3 (May–June 1982), pp. 6–15.

[45] Joseph Cummins and Jerzy Georgiades, "How it Began," *Archivum Immunologiae & Therapie Experimentalis* (1993), Vol. 41, No. 3–4, pp. 169–172.

[46] V.D. Soloviev, "The Results of Controlled Observations of the Prophylaxis of Influenza with Interferon," *Bulletin of the World Health Organization* (1969), Vol. 41, pp. 683–688.

[47] W.A. Tompkins, "Immunomodulation and Therapeutic Effects of the Oral Use of Interferon-*a*," *Journal of Interferon and Cytokine Research*, Vol. 19, No. 8, pp. 817–828 (August 1999).

[48] S.A. Brod, "Autoimmunity is a Type I Interferon Deficiency Syndrome Corrected by Ingested Type I IFN Via the GALT System," *Journal of Interferon and Cytokine Research*, Vol. 19, No. 8, pp. 841–852 (August 1999).

[49] M. Tanaka-Kataoka, T. Kunikata, S. Takayama, et al., "Oral Use of Interferon-*a* Delays the Onset of Insulin-Dependent Diabetes Mellitus in Nonobese Diabetes Mice," *Journal of Interferon and Cytokine Research*, Vol. 19, No. 8, pp. 877–880 (August 1999).

[50] Y.I. Satoh, K. Kasoma, M. Kuwabra, et al., "Suppression of Late Asthmatic Response by Low-Dose Oral Administration of Interferon-*B* in the Guinea Pig Model of Asthma," *Journal of Interferon and Cytokine Research*, Vol. 19, No. 8, pp. 887–894 (August 1999).

[51] M.J. Cummins, "Low-Dose Oral use of Human Interferon-A in Cancer Patients," *Journal of Interferon and Cytokine Research*, Vol. 19, No. 8, pp. 937–940 (August 1999).

[52] Phillip I. Marcus, "Foreword," *Journal of Interferon and Cytokine Research*, Vol. 19, No. 8, p. 813 (August 1999).

[53] Stephen Mamber, Jeremy Lins, Volkan Gurel, et al., "Low-Dose Oral Interferon Modulates Expression of Inflammatory and Auto-Immune Genes in Cattle," *Veterinary Immunology and Immunopathology*, Vol. 172, 64–71 (2016); doi.org/10.1016/j.vetimm.2016.03.006.

[54] Ibid.

Chapter 5

[55] J.A. Georgiades, "Effect of Low-Dose Natural Human Interferon Alpha Given Into the Oral Cavity on the Recovery Time and Death Loss in Feedlot Hospital Pen Cattle: A Field Study," *Arch Immunol Ther Exp* (Warsz): 1993;41(3):205–7.

[56] Joseph M. Cummins, J. Gawthrop, David P. Hutcheson, et al., "The Effect of Low Dose Oral Human Interferon Alpha Therapy on Diarrhea in Veal Calves," *Archivum Immunologiae et Therapie Expermentalis*, Vol. 43 (3–4), January 1993.

[57] Joseph Cummins & D. Hutcheson, "Dose Titration of Human Interferon Alpha Administered Orally for Fever Reduction During Virulent Infectious Bovine Rhinotracheitis Virus Infection," Unpublished 1987: Texas A&M University Ag Research Station.

[58] Joseph Cummins & D. Hutcheson, "The Use of Human Interferon Alpha in Two Hundred Feeder Calves Prior to Shipment from Tennessee to Texas," Unpublished 1984: Texas A&M University Agricultural Research Station.

[59] A.S. Young, A. C. Martin, D.P. Kariuki, et al., "Low-Dose Oral Administration of Human Interferon Alpha Can Control the Development of Theileria Parva Infection in Cattle," *Parasitology*, 101 Pt. 2:201–9. Doi: 10.1017/s0031182000063241, October 1990.

[60] Ibid.

[61] "Treatment of Rotavirus Diarrhea in Calves," Approval Granted by the Ministry of Agriculture, Forestry and Fisheries of Japan, July 2004. *Biovet*, Tokyo, Japan.

[62] Hirochimi Ohtsuka, Mayumi Tokita, Katsushige Takahashi, et al., "Peripheral Mononuclear Cell Response in Japanese Black Calves After Oral Administration of IFN-Alpha," *Journal of Veterinary Medical Science*, 68(10): 1063–7 (October 2006); doi: 10.1292/jvms.68–1063.

[63] Joseph M. Cummins, Mary B. Tomkins, Richard G. Olsen, et al., "Oral Use of Human Alpha-Interferon in Cats," *Journal of Biological Response Modifiers*, Vol. 7, 513–523 (1998).

[64] Ibid.

[65] E. Pedretti, B. Passeri, M. Amadori, et. al, "Low Dose Interferon-Alpha Treatment for Feline Immunodeficiency Virus Infection," *Veterinary Immunology and Immunopathology*, 109: 245–254 (2005): doi: 10.101016/j.vetimm.205.8.020.

[66] Ibid.

[67] Joseph Cummins & David Moore, "Effects of Very Low-Dose Oral Cytokine Treatment on Hematological Values in Feline Leukemia Positive Cats," *Journal of Interferon Research*, 14(1): S187 (1994).

[68] Brian Gilger, Patricia Rose, Michael Davidson, et al., "Low-Dose Oral Administration of Interferon-*a* for the Treatment of Immune-Mediated Keratoconjunctivitis Sicca in Dogs," *Journal of Interferon Research and Cytokine Research*, Vol. 19: 901–905 (1999).

[69] Yoko Ogawa, Eisuke Shimizu & Kazuo Tsubota, "Interferons and Dry Eye in Sjogren's Syndrome," *International Journal of Molecular Science* (November 2019), 10;19(11):3548, doi:10.3390/ijms19113548.

[70] Ian Moore, Barbara Horney, Kendra Day, et al., "Treatment of Inflammatory Airway Disease in Young Standardbreds with Interferon Alpha," *The Canadian Veterinary Journal*, 2004: July 45(7); 594–601.

[71] Ibid.

[72] Bonnie Rush Moore, Steven Krakowka, et al., "Changes in Inflammatory Cell Populations in Standardbred Racehorses after Interferon-Alpha Administration," *Veterinary Immunology and Immunopathology*, 49 (1996) 347–358.

[73] Bonnie Rush Moore, Steven Krakowka, Joseph M. Cummins & James T. Robertson, "Inflammatory Markers in Brochoalveolar Lavage Fluid of Standardbred Racehorses with Inflammatory Airway Disease: Response to Interferon-Alpha," *Equine Veterinary Journal*, Vol. 29 (2) 142–147 (1997).

[74] Joseph Cummins, Richard Mock, Bradford Shive, et al., "Oral Treatment of Transmissible Gastroenteritis with Natural Human Interferon Alpha: A Field Study," *Veterinary Immunology Immunopathology*, 45(3): 355–360 (April 1995): doi: 10.1016/0165–2427(94)05351–R.

[75] Joseph Cummins, Manfred Beilharz, & Steven Krakowka, "Oral Use of Interferon," *Journal of Interferon and Cytokine Research*, 19: 853–857 (1999).

[76] J.G. Leece, J. M. Cummins & A.B. Richards, "Treatment of Rotavirus Infection in Neonate and Weanling Pigs Using Natural Human Interferon Alpha," *Molecular Biotherapy*, Vol. 2(4):211–6 (December 1990).

[77] M. Amadori, P. Candotti, B. Begni & A. Nigrelli, "Oral Treatment of Pigs for PRRS and PMWS," *L'Osservatorio*, Vol. 4:4–5 (2002).

Chapter 6

[78] R.A. Dies, B. Perdereau & E. Falcoff, "From Old Results to New Perspectives: A Look at Interferon's Fate in the Body," *Journal of Interferon Research*, Vol. 7, No. 5, May 4, 2009, doi.org/10.1089/jir.1987.7.553.

[79] Julius A. Lecciones, Norma H. Abejar, Efren E. Dimaano, et al., "A Pilot Double-Blind, Randomized, and Placebo-Controlled Study of Orally Administered IFN-a-n1 (Ins) in Pediatric Patients with Measles," *Journal of Interferon and Cytokine Research*, Vol. 18, pp. 647–652, No. 9, 1998: doi.org/10.1089/jir.1998.18.647.

[80] Eva Perez Martin, Fayna Diaz-San Segundo, Marcelo Weiss, et al., "Type III Interferon Protects Swine Against Foot-and-Mouth Disease," *Journal of Interferon and Cytokine Research*, Vol. 34, No. 10, pp. 810–821, October 2, 2014: doi.org/10.1089/jir.2013.0112.

[81] Rudragouda Channappanavar, Anthony R. Fehr, et al., "IFN-I Response Timing Relative to Virus Replication Determines MERS Coronavirus Infection Outcomes", *Journal of Clinical Investigation*, July 29, 2019: doi.org/10.1172/JCI126363.

[82] Val Hutchison, Juaneve Angenend, William Mok, et al., "Chronic Recurrent Aphthous Stomatitis: Oral Treatment with Low-Dose Interferon Alpha," *Molecular Biotherapy*, Vol. 2 (3): 160–164, August 1990.

[83] Jerzy Georgiades, "Early Changes in the Plasma Proteins of Patients Treated with Low Doses of Oral Natural Human Interferon Alpha, (IFN-a)," *Archivium Immunologiae et Therapie Experimentalis*," Vol. 41, 259–265 (1993).

[84] J. A. Georgiades, J. Caban, T. Zyrkowska-Bieda, et al., "Natural Human Interferon-a Given Orally has Different Effects on Patients with Distinct Forms of Chronic Viral Hepatitis B," *Archivium Immunologiae et Therapie Experimentalis*," Vol. 44 (2–3), 187–194 (1996).

[85] Lee Chuan-Mo, Chi-Yi Chen, Chein Rong Nan, et al., "A Double-Blind Randomized Controlled Study to Evaluate the Efficacy of Low-Dose Oral Interferon-Alpha in Preventing Hepatitis C Relapse," *Journal of Interferon & Cytokine Research*, Vol. 34, No. 3, p. 187-194: doi: doi.org/10.1089/jir.2013.0074.

[86] Bill Miller, "Interferon Enters the Fray," *Farm Journal*, October 1985, pp. 12–13.

[87] Ibid.

[88] Ibid.

[89] Ibid.

[90] History of Hayashibara Biochemical Laboratories, www.hayashibara.co.jp/data/1283/en_groupinformation_tp/, accessed May 16, 2020.

Chapter 7

[91] Val Hutchison & J. M. Cummins, "Low-Dose Interferon in Patient with Aids," *The Lancet*, Vol. 330, Issue 8574, p. 1530-1531, (December 26, 1987): doi: doi.org/10.1016/S0140-6736(87)92671-7.

[92] Ibid.

[93] New Zealand AIDS Memorial Quilt, "Dr. Buddy Brandt," accessed May 11, 2020, www.aidsquilt.org.nz/dr-buddy-brandt/.

[94] *Parasitology* Vol. 101, pp. 201–209, 1990.

[95] Davy Koech, Arthur Obel, Jun Minowada, et al., "Low Dose Oral Alpha-Interferon Therapy for Patients Seropositive for Human Immunodeficiency Virus Type-1 (HIV-1)," *Molecular Biotherapy*, Vol. 2, June 1990, pp. 91–95.

[96] Ibid., 93.

[97] Ibid., 94.

[98] Ibid.
[99] Gina Kolata, "Ignored AIDS Drug Shows Promise in Small Tests," *New York Times*, August 15, 1989.
[100] Ibid.
[101] Ibid.
[102] Ibid.
[103] Ibid.
[104] Ibid.
[105] "Experienced Smuggler Says AIDS Drug Worth Risk," *San Antonio Light*, September 16, 1990.
[106] Ibid.
[107] Ibid.
[108] Ibid.
[109] Ibid.

Chapter 8

[110] Max Albright, "AIDS Fighter–Amarillo Veterinarian Awaits Tests of His Discovery," *Amarillo Sunday News–Globe*, April 15, 1990.
[111] Kimerleigh J. Smith, "U.S. Patent Sought for Controversial AIDS Drug," *The City Sun*, February 6–12, 1991.
[112] Ibid.
[113] Ibid.
[114] Christopher Anderson, "Racial Tensions Entangle NIH in Dispute Over AIDS Drug," *Nature*, October 22, 1992, pp. 660–661.
[115] Ibid., 661.
[116] Ibid.
[117] Michael Hutton, David Levin & Laurence Freedman, "Randomized, Placebo-Controlled, Double-Blind Study of low-Dose Oral Interferon-a in HIV-1 Antibody Positive Patients," *Journal of Acquired Immune Deficiency Syndromes*, 5: 1084–1090 (1992).
[118] G. Kaiser, H. Jaeger, J. Birkman, et al., "Low-Dose Natural Human Interferon-Alpha in 29 patients with HIV-1 Infection: A Double-Blind, Randomized, Placebo Controlled Trial," *AIDS*, June 6, 1992, pp. 563–569: doi: 10.1097/00002030–19926000 –00007.
[119] "The Many Types and Health Benefits of Kale," Mayo Clinic Health System, www.mayoclinichealthsystem.org/hometown-health/speaking-of-health/the-many -types-and-health-benefits-of-kale (Accessed May 27, 2020).
[120] P. Thongcharoen, C. Wasi & S. Sarasombath, "Low-Dose Oral Interferon Therapy of Asymptomatic and Symptomatic HIV-1 Seropositive Individuals," January 1992. Studies\hiv\thailand.595-051795pab.
[121] M. Mukunyandela, A.S. Richards & M.J. Cummins, "Treatment of Symptomatic HIV-1 Infected Patients with Low-Dose Oral Natural Human Interferon Alpha," *Journal of Interferon Research*, 14(1):S191, 1994.
[122] Beverly Alston, Jonas Ellenberg, Harold Standiford, et al., "A Multicenter, Randomized, Controlled Trial of Three Preparations of Low Dose Oral a-Interferon

in HIV-Infected Patients with CD4+ Counts Between 50 and 350 cells/mm3," *Journal of Acquired Immune Deficiency Syndromes*, 22:348–357 (August 18, 1999).
[123] Ibid.
[124] E. Katabira, N. Sewankamo, R. Mugerwa, et al., "Lack of Efficacy of Low-Dose Oral Interferon Alfa in Symptomatic H& IV-1 Infection: A Randomized, Double-Blind, Placebo Controlled Trial," *Sex Transm Inf.*, 74:265–270 (1998).
[125] Geoffrey Cowley & Mary Hager, "The Angry Politics of Kemron," *Newsweek*, January 4, 1993.
[126] Ibid.
[127] William Gaines & David Jackson, "AIDS Hope or Hoax in a Bottle?" *Chicago Tribune*, March 14, 1995.
[128] Ibid.
[129] Wilbert Jordan, "Three Open Label Studies of Oral Interferon Alpha in the Treatment of HIV Disease," *Journal of the National Medical Association*, Vol. 86, No. 4, pp. 257–262 (1994).
[130] Joseph Cummins, "Clinical Studies in HIV+ Patients Given Low-Dose Oral Interferon (IFN)," *Current Research on HIV/AIDS*, Vol. 2018; Issue 1, pp. 1–4: doi: 10.29011/2575-7105/100017.
[131] S.E. Wright, D.P. Hutcheson & J.M. Cummins, "Low dose Oral Interferon Alpha 2a in HIV-1 Seropositive Patients: A Double-blind, Placebo-controlled Trial," *Biotherapy*, 11:229–234, 1998.

Chapter 9

[132] "Atlas of the Atmosphere," *The Scientist*, December 1, 2010.
[133] Arthur Ericsson & Almer Cabler Engle, III, "Clinical Studies Phase II: Polymyositis," *Raum & Zelt*, Vol. 2, No. 6, pp. 10–14 (1991).
[134] Arthur Ericson & Joseph Cummins, "Clinical Studies of Neuromuscular Disease," *Raum & Zelt*, Vol. 2, No. 5, pp. 26–32 (1991).
[135] Ibid., p. 28.
[136] Ibid.
[137] T.R. Benkendorfer, Arthur Ericsson, F.G. Kigadye, et al., "Acquired Immunodeficiencey Syndrome Treated with VIRON," *Explore!*, Vol. 3, No. 4, pp. 9–13 (1992).
[138] Staley Brod, "Type I Interferon: A Potential Treatment for Autoimmunity," *Journal of Interferon and Cytokine Research*, 22(12): 1153–66, (January 2003): doi: 10.1089/10799900260475669.
[139] Martin Cummins & Brian Pruitt, "Low-Dose Oral Use of Human Interferon-Alpha in Cancer Patients," *Journal of Interferon and Cytokine Research,* Vol. 19, No. 8 (July 7, 2004): doi 10.1089/107999099313488.
[140] F. Leveque, M Al-Saraff, J. Kish, et al., "Low-Dose Oral Human Interferon Alpha (HuIFNa) to Treat Muscositis Induced by Chemotherapy." Presented at American Association for Cancer Research Annual Meeting, March 28–April 1, 1998, New Orleans, LA.
[141] Joseph Cummins & Jerzy Georgiades, "How it Began," *Archivum Immunologiae & Therapiae Experimentalis*, Vol. 41, No. 3–4, pp. 169–172 (1993).

[142] M.J. Cummins, A. Pappas & P.C. Fox, "Treatment of Primary Sjogren's Syndrome with Low-Dose Human Interferon Alfa Administered by the Oromucosal Route: Combined Phase II Results," *Arthritis and Rheumatism*, 49(4):583–593: doi: 10.1002/art.1199.

[143] M.J. Cummins, A. Pappas, G.M. Kammer & P.C. Fox, "Treatment of Primary Sjogren's Syndrome with Low-Dose Human Interferon Alfa Administered by the Oromucosal Route: Combined Phase III Results," *Arthritis and Rheumatism*, 49(4):583–593 (July 31, 2003): doi: 10.1002/art.1199.

[144] Jon Russel, Joel Michalek, Yoon-Kyo Kang & Alan Richards, "Reduction of Morning Stiffness in Physical Function in Fibromyalgia Syndrome Patients Treated Sublingually with Low Doses of Human Interferon-A," *Journal of Interferon and Cytokine Research*, 19:961–968 (1999).

[145] D. Greenspan, L. Macphal, B. Cheikh, et al., "Low-Dose Interferon-Alpha (IFNa) in the Treatment of Oral Warts in HIV Patients," *Journal of Dental Research*, (Special Issue): 187 (#216), January 2001.

[146] Lorenzo Lutherer, Kenneth Nugent, Byron Schoettle, et al., "Low-Dose Interferon A Possibly Retards the Progression of Idiopathic Pulmonary Fibrosis and Alleviates Associated Cough in Some Patients," *Thorax*, Vol. 66, Issue 5, pp. 446–447 (2011).

[147] Cummins et al., *Am J Vet Res*, 66[1]:164–176, 2005.

[148] Georgiades, *Arch Immunol Ther Exp*, 41[3–4]:205–207, 1993.

[149] Tovey et al., Abstract 137, *Ann Mtg ISICR*, 2003; Dron et al.; *Genomics* 79:315–325, 2002; and Namangula et al., *JICR* 26:675–681, 2006.

[150] Jack Gwaltney, University of Virginia, Studies 1993–1997.

Chapter 10

[151] "Common Human Coronaviruses," Centers for Disease Control and Prevention (Last reviewed February 13, 2020) (Accessed July 6, 2020), www.cdc.gov/coronavirus /general-information.html.

[152] Abigail Smith, Stephen Barthold & Deborah Beck, "Intranasally Administered Alpha/Beta Interferon Prevents Extension of Mouse Hepatitis Virus, Strain JHM, Into the Brains of BALB/cByJ Mice," *Antiviral Research*, Vol. 8, pp. 239–246 (October 23, 1987).

[153] Ibid., 223–224.

[154] Joseph Cummins, Richard Mock, Bradford Shive, et al., "Oral Treatment of Transmissible Gastroenteritis with Natural Human Interferon Alpha: A Field Study," *Veterinary Immunology and Immunopathology*, Vol. 45, pp. 355–369 (June 1994).

[155] Mitch Leslie, "T Cells Found in COVID-19 Patients 'Bode Well' for Long-Term Immunity," *Science* magazine, May 14, 2020, www.sciencemag.org/news/2020/05 /t-cells-found-covid-19-patients-bode-well-long-term-immunity.

[156] Ibid.

[157] Ibid.

[158] Emily Mantlo, Natalya Bukreyeva, Junki Maruyama, et al., "Antiviral Activities of Type I interferons to SARS-CoV-2 Infection," *Journal of Antiviral Resistance*, Vol. 178, 104811 (April 29, 2020): doi: 10.1016/j.antiviral.2020.104811.

[159] Ibid.

[160] Ibid.

[161] Ivan Fan Hung, Kwok-Cheung Lung, et al., "Triple Combination of Interferon Beta-1b, Lopinavir-Ritonavir, and Ribavirin in the Treatment of Patients Admitted to Hospital with COVID-19: An Open-Label, Randomized, Phase 2 Trial," *The Lancet*, Vol. 395: 1695–1704: doi: 10.1016/S0140–6736(20)31042–4.

[162] Effat Davoudi-Monfared, Hamid Rahmani, Hossein Khalili, et al., "Efficacy and Safety of Interferon Beta-1a in Treatment of Severe COVID-19: A Randomized Clinical Trial," *MedRxiv*, May 28, 2020, www.medrxiv.org/content/10.1101/2020.05.28.20116467v1, doi: 10.1101/2020.05.28.20116467.

[163] Qiong Zhou, Virginia Chen, et al., "Interferon-*a*2b Treatment for COVID-19," *Frontiers in Immunology*, Vol. 11, Article 1601, May 15, 2020: doi: 10.3389/fimmu.2020.01061.

[164] Ibid.

[165] Ibid.

ACKNOWLEDGMENTS

It took lots of people and money (more the $40 million) to help me follow my passion to study interferon for fifty years. Some of the people I recall herein supported me with their time, money, and encouragement. I regret I was unable to repay these folks with FDA approval or recognition of the value of the technology. Even with help from Mitsubishi and Hayashibara, I simply was not smart enough to negotiate all the pitfalls. Some people are mentioned by name in the book. Others are named below:

Ed Amento, Karen and Bill Barnett, David Bechtol, Manfred Beilharz, Gail Berger, Steve Berk, Burton Berkson, Randy Berrier, Tom Bivins, Charles Black, Kathy Brecht, Casey Burns, Patty Caldwell, Barry Carter, Steve Chen, Dorothy and Bill Clymer, Bernard Cohen, Charlie Cobb, Jim Cook, Tim Cunningham, Tom D'Alonzo, Annie Ewing, Rachel Fields, Robert Fischer, Bob Fleischmann, Phil Fox, Manasori Fugii, Joseph Garofalo, Jan Gawthrop, Gary Graham, Dan Guthrie, Rolf Habersang, Katsuaki Hayashibara, Ken Hayashibara, Joe Huff, Charles Hughes, Bill Jongsma, Kathleen Kelleher, Hiroshi Kojima, Steve Krakowka, Masashi Kurimoto, Steve Lammert, Michael Lange, Fran Leveque, Tyrrell Levitt, Allan Lieberman, Lorenz Lutherer, Ray Loan, Steve Mamber, Claus Martin, John McMichaels, Don McTaggert, Marv Pflaumer, Joel Michalek, Kim Miller, Dave Min, Richard Mock, Ed Morris, Phillippe Niemetz, Hy Ochberg, Kunihiro Ohashi, Jim Page, Don Paxton, Wally Raubenheimer, Sam Reeves, Alan Richards, Jake Richards, Ken Ridenour, Sherry Risinger, Bruce Rosenquist, Whitney Rounds, Bonnie Rush, I. Jon Russell, Fred Sattler, George Segel, Crystal Shelton, Ed Sherwood, Shunichi Shiozawa, Jonathan Ship, John Smith, Kelly Smith, Michelle Snustead, David Stewart, Doug Testa, Mary and Wayne Tompkins, GB Thomson, Paul Tibbits, Dot Tutt, Tom Ulie, Bill Varner, Shannon Warren, Billy Walters, Steve Whitney, Bryan Williams, Lemmy Wilson, Roger Wyatt, Hasan Zeytin.

I am of Irish and German descent. I am foolish, careless, selfish, and stubborn. My steadfastness in working all the time was hard on my family. I am blessed that Ella Rose has stuck with me these past twenty-seven years. The reason I know God exists is because He sent the angel Ella Rose to care for me.

—Dr. Joseph Cummins

I'd like to thank my lovely wife and partner in life, Linda, my son and first reader, Ben, and my daughter Jacqueline who shows me on a daily basis the meaning of courage.

I'm appreciative for my father, who has always supported my efforts and my dear, deceased mother who taught me that if a cause is just you fight no matter the size of your enemy. I'd like to thank my brother, Jay, who has been my champion and best friend since the day I was born.

I'd like to especially thank some of the great teachers in my life, my seventh grade science teacher, Paul Rago, high school teachers, Ed Balsdon and Brother Richard Orona, college professors Clinton Bond, David Alvarez, and Carol Lashoff, writing teachers James Frey, Donna Levin, and James Dalessandro and in law school, the always entertaining Bernie Segal.

I've been blessed to have some of the greatest friends anybody could ever want who have stood by me through both success and failure. Thank you, John Wible, John Henry, Chris Sweeney, Pete Klenow, Suzanne Golibart, Beth Bergen, Jyoti Dave, Sue Brown, Eric Holm, Sherilyn Todd, Bill Wright, and Max Swafford for your constant friendship.

I want to thank some of my great science colleagues at Gale Ranch Middle School, Matt Lundberg, Danielle Pisa, Derek Augarten, Amelia Larson, Arash Pakdel, and Katie Strube.

And last I'd like to thank the wonderful staff at Skyhorse Publishing, the fabulous Caroline Russomanno and publisher Tony Lyons for your continuing faith in me.

—Kent Heckenlively, JD